Analysis of unbalanced data

A pre-program introduction

Analysis of unbalanced data
A pre-program introduction

CHING CHUN LI

University Professor of Biometry and Human Genetics
Graduate School of Public Health, University of Pittsburgh

CAMBRIDGE UNIVERSITY PRESS

CAMBRIDGE

LONDON NEW YORK NEW ROCHELLE

MELBOURNE SYDNEY

Published by the Press Syndicate of the University of Cambridge
The Pitt Building, Trumpington Street, Cambridge CB2 1RP
32 East 57th Street, New York, NY 10022, USA
296 Beaconsfield Parade, Middle Park, Melbourne 3206, Australia

First published 1982

Printed in the United States of America

Library of Congress catalogue card number: 82-4253

British Library cataloguing in publication data

Li, Ching Chun
Analysis of unbalanced data.
1. Probabilities 2. Statistical mathematics
I. Title
519.2 QA273

ISBN 0 521 24749 7

AS

To the memory of
Professor Antonio Ciocco
1908–1972

Contents

Preface

One of the frequently encountered problems in my experience of statistical consultation is the analysis of unbalanced two-way classification data. This type of data arises from practically every field of research: biology, epidemiology, medicine, as well as the social sciences. For orthogonal data such as those of randomized blocks, we can give the research worker an arithmetic procedure (a recipe, if you will) which can be followed in an easy routine manner. But not so for unbalanced data.

I used to spend long hours with the research worker, going through the analysis of the unbalanced data. My experience shows that the statistical procedure proves more confusing than enlightening if no reasonable explanation is offered to the research worker. In other words, some understanding is necessary in order to follow the statistical procedure intelligently.

In the mid-1950s, Professor Antonio Ciocco, then the Chairman of the Department of Biostatistics at the Graduate School of Public Health, University of Pittsburgh, sizing up the situation, suggested that I give a short course devoted exclusively to the analysis of unbalanced data for general research workers and students of applied statistics. Thus, a modular course on statistical methods was born. The present volume is an outgrowth of the notes from that modular course.

The exposition is by way of numerical examples; the explanations are in terms of the most elementary algebra. No mathematical proficiency is required. Research workers who have had an introductory course in statistics up to one-way classification will have no difficulty in learning at least this much of the methods dealing with unbalanced data.

The present volume may be used for self study or as a text for a modular course in statistical methods. In my school I cover the material (not including the Appendices) in six or seven lectures as a pre-program introduction to the analysis of unbalanced data. Some of my students are also taking a more mathematical course in linear models, but they seem to like the arithmetic exposition just as well as those students not taking the mathematical course.

There is some repetition here and there in the book. One reason for the redundancy is to make the various chapters more or less self-contained so that a reader may consult a particular chapter or topic without going through the whole book. But there may be some redundancy that cannot be justified on that ground. If so, I apologize – sincerely but not too profusely. Research in education discovered that reinforcement (a modern term for repetition) is a powerful tool with which to learn.

One final word to the student who studies the book by himself. One cannot just read the book; he should have pencil, paper, and perhaps a pocket calculator to work through the examples, verify the various relationships, and do the exercises at the end of each chapter. If he does all these, he will not only find the book easy to follow, but he will also gain some understanding of the methods.

C. C. Li

1

Linear equations

The only mathematical knowledge required to work through the computing procedure of analysis, as presented in this book, is the method of solving a set of linear equations. At a certain stage of the statistical analysis we shall encounter the problem of solving linear equations. Perhaps it is helpful to review the method of solving linear equations at the outset, so that our description of the analysis in subsequent chapters will not be interrupted by excessive arithmetic. The type of linear equations that occur in statistical analysis may be exemplified by three equations with three unknowns:

$$\left.\begin{aligned} \tfrac{13}{6}x - \tfrac{5}{6}y - \tfrac{4}{3}z &= -101 \\ -\tfrac{5}{6}x + \tfrac{23}{12}y - \tfrac{13}{12}z &= 58 \\ 4x + 3y + 4z &= 0 \end{aligned}\right\} \qquad (1)$$

Clearing the fractions,

$$\left.\begin{aligned} 13x - 5y - 8z &= -606 \\ -10x + 23y - 13z &= 696 \\ 4x + 3y + 4z &= 0 \end{aligned}\right\} \qquad (1')$$

This set of equations may be solved by the method we learned in high school, without bother or fuss, as our only purpose is to find the numerical values of (x, y, z) that satisfy the equations.

Those who are familiar with determinants may use Cramer's rule; and those who are familiar with matrix algebra may use the inverse matrix. Here we shall use the high school method to illustrate one aspect of my attitude towards a modular course: that is, the principle of minimum learning. If we use any method beyond the high school level, the reader will have to acquaint himself with that method and that necessitates

extra learning. Our purpose is to learn the arithmetic of statistical analysis of two-way classification data; and we like to learn it the shortest way possible, involving a minimum of extra detours.

The high school method calls for making the coefficients of one of the unknowns equal, so that that unknown may be eliminated by subtraction, resulting in two equations with two unknowns. Let us eliminate x first. Multiplying the second equation by 1.30 and the third equation by $\frac{13}{4} = 3.25$, we obtain

$$\left.\begin{array}{lr} \text{(i)} & 13.00x - 5.00y - 8.00z = -606.00 \\ \text{(ii)} & -13.00x + 29.90y - 16.90z = 904.80 \\ \text{(iii)} & 13.00x + 9.75y + 13.00z = 0 \end{array}\right\} \tag{1''}$$

so that (i) + (ii) and (iii) − (i) yield two equations with y and z only:

$$\text{(iv)}\quad 24.90y - 24.90z = 298.80$$

$$\text{(v)}\quad 14.75y + 21.00z = 606.00$$

Equation (iv) yields $y - z = 12$. Substituting $y = 12 + z$ in (v), we obtain one equation with only one unknown:

$$14.75(12 + z) + 21z = 606$$

$$35.75z = 429$$

giving $z = 12$, and $y = 12 + z = 24$. Substituting these in any one of the original three equations, we may solve for x. Obviously, the third equation $(4x + 3y + 4z = 0)$ is the most convenient to use. Thus, $4x = -3y - 4z = -3(24) - 4(12) = -120$ so that $x = -30$. Summary: the solution of the equations is

$$(x, y, z) = (-30, 24, 12) \tag{2}$$

This solution is unique and there are no other solutions.

One of the nice things about solving equations is that we can always check the answer, so that there is no excuse for having wrong solutions. The reader should take time to see that the solution (2) does satisfy the equation (1) or its equivalents (1′) and (1″).

In the numerical example above, we have eliminated the unknown x first and then we proceed to solve the two equations involving y and z. As an exercise, the reader may proceed otherwise, such as eliminate z first and then solve the two equations involving x and y. In any case, the solution remains the same, as we have learned in high school.

The simple remarks above may be slightly generalized. Suppose there are five equations with five unknowns $(x_1, x_2, y_1, y_2, y_3)$. In the process of solving these equations we may eliminate the three ys first and then

solve the two equations for x_1 and x_2. Or, we may eliminate the two *x*s first and then solve the three equations for the *y*s. In either case, the final solution for the five unknowns remains the same. We shall have occasion to make use of this principle in statistical analysis.

Two equations with three unknowns

A general discussion of the situation of having more unknowns than equations is beyond the scope of this course. In statistical analysis of two-way classification data we shall encounter the case in which the number of equations is one (or two) less than the number of unknowns. Hence, we consider the case of two equations with three unknowns as an illustration. To be concrete, let us consider the first two equations of (1) or, more conveniently, the first two equations of (1″), assuming the third equation non-existent. In order to obtain a solution, we may regard one of the unknowns as an arbitrary number and transfer it to the right-hand side of the equations. Suppose we regard z as an arbitrary number. Then the first two equations of (1″) may be rewritten as

$$\left.\begin{aligned} 13.0x - 5.0y &= -606.0 + 8.0z \\ -13.0x + 29.9y &= 904.8 + 16.9z \end{aligned}\right\} \qquad (3)$$

which are to be regarded as two equations in two unknowns (x, y), since z is now an arbitrary number. Eliminating x by adding these two equations, we obtain

$$\left.\begin{aligned} 24.9y &= 298.8 + 24.9z \\ y &= 12 + z \\ &\text{(substituting in (3)),} \\ x &= -42 + z \end{aligned}\right\} \qquad (4)$$

The solution (4) is called the *general solution* of equation (3). We see that there is not just one solution but an infinite number of solutions, depending on the arbitrary number z. For example, with $z = 0, 7, 12$, the respective solutions are

$$(x, y, z) = (-42, 12, 0), (-35, 19, 7), (-30, 24, 12)$$

Again, the reader should take time to check that each of these solutions satisfies the two equations in (3). Needless to say, no solution is 'better' than the others. They are all bona fide solutions. Also observe that, when $z = 12$, the solution $(-30, 24, 12)$ is the same as the unique solution (2) of equation (1). Thus, the general solution (4) of equation (3) includes the unique solution (2) of equation (1).

Independent and dependent equations

The three equations in (1), permitting a unique solution (2), are known as linearly independent equations, because each equation states an independent condition that the unknowns (x, y, z) must satisfy. Not all sets of three equations permit us to find unique solutions. Consider the following, for example,

$$\left.\begin{array}{ll}\text{(i)} & x+3y-2z=10 \\ \text{(ii)} & 2x-\ \ y+3z=13 \\ \text{(iii)} & 3x+2y+\ \ z=23\end{array}\right\} \qquad (5)$$

where the third equation is simply the sum of the preceding two. Then any solution that satisfies (i) and (ii) will automatically satisfy (iii) also. The third equation sets no extra and independent conditions on the unknowns; it is merely a repetition of the conditions already set by (i) and (ii). Hence, the third equation is useless to us.

Since (iii) = (i) + (ii), we may also write (ii) = (iii) − (i) and regard the second equation as dependent on the first and third. Similarly, we may regard (i) = (iii) − (ii). In other words, there are only two independent equations. Any two of the three equations of (5) are independent and the remaining one dependent (a repetition). Under these circumstances, there is no use in us pretending to have three equations, and there is no possibility of finding a unique solution. What we should do is to treat the case as in (3), two equations with three unknowns, finding a solution of the type (4). For the particular example (5), it is good exercise for the reader to verify the general solution:

$$x=7-z, \quad y=1+z \qquad (6)$$

In the above we let (iii) = (i) + (ii) for simplicity. In fact, if (iii) is any linear combination of (i) and (ii), what we said above still holds. For instance, let (iii) = k_1(i) + k_2(ii), where k_1 and k_2 are constants. The fact remains that any solution that satisfies (i) and (ii) will automatically satisfy (iii) also. The third equation confers no independent conditions on the unknowns and is useless to us.

There are methods to detect if equations are linearly independent or not, but that is a subject beyond the scope of this book. Even the rudimentary knowledge of linear equations outlined in this section will be helpful to us in analyzing two-way classification data. Other properties of the equations will be discussed when they occur in later sections.

Exercises

A necessary step in analyzing unbalanced two-way classification data is to solve a set of linear equations. Those who have not solved equations lately may need some practice with simple examples.

1. Simplest example: two equations with two unknowns.

$$\tfrac{13}{6}x - \tfrac{5}{6}y = -101$$
$$-\tfrac{5}{6}x + \tfrac{23}{12}y = \quad 58$$

Hint: Clear the fractions first.

Answer: $(x, y) = (-42, 12)$.

2. Still two equations with two unknowns; c is a constant.

$$-\tfrac{13}{6}c + \tfrac{5}{6}y + \tfrac{4}{3}z = 101$$
$$-\tfrac{5}{6}c + \tfrac{23}{12}y - \tfrac{13}{12}z = \quad 58$$

Hint: Clear the fractions and transpose the terms involving c to the right-hand side of the equations and then solve for y and z.

Answer: $(y, z) = (54 + c, 42 + c)$.

3. Solve the following three equations:

$$-\tfrac{10}{12}x + \tfrac{23}{12}y - \tfrac{13}{12}z = 58$$
$$-\tfrac{16}{12}x - \tfrac{13}{12}y + \tfrac{29}{12}z = 43$$
$$4x + 3y + 4z = \quad 0$$

Answer: $(x, y, z) = (-30, 24, 12)$, which is the same as the solution of equation (1) in the text. Note that the second equation here is the sum of the first two equations of (1), with sign changed.

4. Solve the following three equations:

$$13x - \quad 5y - \quad 8z = -606$$
$$-10x + 23y - 13z = \quad 696$$
$$x + \quad y + \quad z = \quad 0$$

Hint: One method is to eliminate z by substituting $z = -x - y$ in the first two equations.

Answer: $(x, y, z) = (-32, 22, 10)$.

2

Quadratic forms

To facilitate the recognition of certain expressions in our later analysis, it is helpful to get acquainted with quadratic forms at this early stage. Here, we shall only deal with the way of writing a quadratic form, not with its theory, which is a wide topic in mathematics and statistics. Algebraic expressions like

$$\phi = a_{11}x_1^2 + a_{22}x_2^2 + 2a_{12}x_1x_2$$

and

$$\phi = a_{11}x_1^2 + a_{22}x_2^2 + a_{33}x_3^2 + 2a_{12}x_1x_2 + 2a_{13}x_1x_3 + 2a_{23}x_2x_3 \quad (1)$$

are called quadratic forms, where the *a*s are numerical coefficients and the *x*s are variables. Thus, a quadratic form is a function of the *x*s. When the *x*s are assigned certain values the quadratic form ϕ will assume a certain value which may be positive, zero, or negative, depending on the coefficients (*a*s) and the particular values the *x*s assume.

The coefficient of a square term, x_i^2, is a_{ii}. That the coefficient of a product term, x_ix_j, is written as $2a_{ij}$ should not bother the reader. When there is a term like $7x_1x_2$, we can always regard it as $2 \times 3.5x_1x_2$. A quadratic form may be conveniently written by arranging the terms systematically into a square array. For example, the quadratic form (1) may be written

$$\begin{aligned}\phi &= \phi(x_1, x_2, x_3)\\ &= \quad a_{11}x_1^2 \quad + a_{12}x_1x_2 + a_{13}x_1x_3\\ &\quad + a_{12}x_1x_2 + a_{22}x_2^2 \quad + a_{23}x_2x_3\\ &\quad + a_{13}x_1x_3 + a_{23}x_2x_3 + a_{33}x_3^2 \end{aligned} \quad (2)$$

In such a 3×3 arrangement, the square terms appear in the principal diagonal and the product terms are symmetrically situated off the

diagonal. Indeed, when the arrangement of a square array of terms is universally agreed upon and understood, we may write only the nine coefficients according to the arrangement pattern of (2), without the explicit writing of the *x*s, for the position of the *a*s in the square array tells us which term it is. Since we do not require knowledge of matrix algebra in reading this text, this is all we say about quadratic forms in general.

Sum of squares as a quadratic form

Now we shall study a particular type of quadratic form which is frequently encountered in statistics, particularly in the analysis of variance which involves the addition or subdivision of sum of squares (of deviations from the mean). Consider n numbers $(x_1, x_2, \ldots, x_n)$ with mean value $\bar{x} = \sum x_i/n$. The sum of squares (of deviations from the mean) is

$$ssq = \sum (x_i - \bar{x})^2 = \sum x_i^2 - \frac{(\sum x_i)^2}{n} \qquad (3)$$

a fundamental algebraic identity proved in all elementary statistics textbooks. The sum of squares (3) may be written as a quadratic form. For those who encounter the new way of writing a sum of squares we begin with only two numbers, x_1 and x_2. The sum of squares of these two numbers is, by (3),

$$\begin{aligned} ssq &= x_1^2 + x_2^2 - \tfrac{1}{2}(x_1 + x_2)^2 = x_1^2 + x_2^2 - \tfrac{1}{2}(x_1^2 + x_2^2 + 2x_1x_2) \\ &= \tfrac{1}{2}x_1^2 - \tfrac{1}{2}x_1x_2 \\ &\quad - \tfrac{1}{2}x_1x_2 + \tfrac{1}{2}x_2^2 \end{aligned} \qquad (4)$$

which is a quadratic form. The sum of squares (4) for two numbers may thus be written, as is well known,

$$ssq = \tfrac{1}{2}(x_1^2 - 2x_1x_2 + x_2^2) = \tfrac{1}{2}(x_1 - x_2)^2 \qquad (4')$$

Consider three numbers (x_1, x_2, x_3). The sum of squares for these three numbers is, by (3),

$$\begin{aligned} ssq &= x_1^2 + x_2^2 + x_3^2 - \tfrac{1}{3}(x_1 + x_2 + x_3)^2 \\ &= x_1^2 + x_2^2 + x_3^2 - \tfrac{1}{3}(x_1^2 + x_2^2 + x_3^2 + 2x_1x_2 + 2x_1x_3 + 2x_2x_3) \\ &= \qquad \tfrac{2}{3}x_1^2 - \tfrac{1}{3}x_1x_2 - \tfrac{1}{3}x_1x_3 \\ &\qquad - \tfrac{1}{3}x_1x_2 + \tfrac{2}{3}x_2^2 - \tfrac{1}{3}x_2x_3 \\ &\qquad - \tfrac{1}{3}x_1x_3 - \tfrac{1}{3}x_2x_3 + \tfrac{2}{3}x_3^2 \end{aligned} \qquad (5)$$

which, again, is a quadratic form. The two quadratic forms, (4) and (5), share many common properties, as both are the sum of squares.

Obviously, the value of such a quadratic form $\phi = ssq$ is always positive, no matter what are the values of x_1, x_2, x_3, except in the trivial case $x_1 = x_2 = x_3$ with $ssq = 0$. It can never be negative. The coefficients of the terms of such quadratic forms also have certain simple properties. In both (4) and (5), the square terms in the diagonal are positive, and the off-diagonal product terms are negative. The coefficients of each row of the square array of terms add up to zero. Due to the symmetrical arrangement of the coefficients, they also add up to zero in each column. The conclusion is that whenever we see a quadratic form with these properties, we know it represents a sum of squares.

Duplicated numbers

We may generalize the results of the previous section by introducing duplicated numbers. As the simplest example of this nature, let us consider the *four* numbers (x_1, x_1, x_2, x_3), in which x_1 appears twice. However, there are only three distinct numbers. The sum of squares for these four numbers is, by (3),

$$\begin{aligned} ssq &= 2x_1^2 + x_2^2 + x_3^2 - \tfrac{1}{4}(2x_1 + x_2 + x_3)^2 \\ &= 2x_1^2 + x_2^2 + x_3^2 - \tfrac{1}{4}(4x_1^2 + x_2^2 + x_3^2 + 4x_1x_2 + 4x_1x_3 + 2x_2x_3) \\ &= \quad x_1^2 - \tfrac{1}{2}x_1x_2 - \tfrac{1}{2}x_1x_3 \\ &\quad - \tfrac{1}{2}x_1x_2 + \tfrac{3}{4}x_2^2 - \tfrac{1}{4}x_2x_3 \\ &\quad - \tfrac{1}{2}x_1x_3 - \tfrac{1}{4}x_2x_3 + \tfrac{3}{4}x_3^2 \end{aligned} \tag{6}$$

which, we recognize, is not only a quadratic form but it represents a sum of squares with coefficients adding up to zero in each row and in each column. The diagonal square terms are positive, and all off-diagonal product terms are negative.

Now, we have seen that a quadratic form is a more general way of writing a sum of squares, taking care of duplicated *x*s.

The interested reader may extend the writing to the case of six numbers, say $(x_1, x_1, x_2, x_3, x_3, x_3)$ in which x_1 appears twice, x_2 once, and x_3 thrice. But, for the sake of arithmetic brevity, we shall only use the simplest example for illustration.

Sum of quadratic forms

Using quadratic forms to denote a sum of squares has other advantages. For instance, it enables us to obtain the total sum of squares of several groups of the same *x*s in one expression. To illustrate the

idea, let us consider the following four groups of *xs*:

	Groups	*Sum of squares* (3)
I	(x_1, x_1, x_2, x_3)	$2x_1^2+x_2^2+x_3^2-\frac{1}{4}(2x_1+x_2+x_3)^2$
II	(x_1, x_3)	$x_1^2+x_3^2-\frac{1}{2}(x_1+x_3)^2$
III	(x_1, x_2, x_3)	$x_1^2+x_2^2+x_3^2-\frac{1}{3}(x_1+x_2+x_3)^2$
IV	(x_2, x_3)	$x_2^2+x_3^2-\frac{1}{2}(x_2+x_3)^2$

(7)

Each of the sum of squares shown above is the sum of squares within that group. We now wish to obtain the total of the four sums of squares shown above. This is easily done by expressing each sum of squares as a quadratic form and then adding the four quadratic forms, which amounts to adding the corresponding coefficients of the four quadratic forms. The quadratic forms for the four sums of squares have already been given in previous sections. For brevity, we merely indicate their coefficients in the following:

$$\begin{array}{cccc} \textit{Group I} & \textit{Group II} & \textit{Group III} & \textit{Group IV} \\ \begin{pmatrix} 1 & -\frac{1}{2} & -\frac{1}{2} \\ -\frac{1}{2} & \frac{3}{4} & -\frac{1}{4} \\ -\frac{1}{2} & -\frac{1}{4} & \frac{3}{4} \end{pmatrix} & \begin{pmatrix} \frac{1}{2} & 0 & -\frac{1}{2} \\ 0 & 0 & 0 \\ -\frac{1}{2} & 0 & \frac{1}{2} \end{pmatrix} & \begin{pmatrix} \frac{2}{3} & -\frac{1}{3} & -\frac{1}{3} \\ -\frac{1}{3} & \frac{2}{3} & -\frac{1}{3} \\ -\frac{1}{3} & -\frac{1}{3} & \frac{2}{3} \end{pmatrix} & \begin{pmatrix} 0 & 0 & 0 \\ 0 & \frac{1}{2} & -\frac{1}{2} \\ 0 & -\frac{1}{2} & \frac{1}{2} \end{pmatrix} \end{array}$$

The sum of the four quadratic forms is obtained by adding the four sets of coefficients together. Thus, for the three elements in the first row (or first column),

$$a_{11} = 1+\tfrac{1}{2}+\tfrac{2}{3}+0 = \tfrac{13}{6}$$
$$a_{12} = -\tfrac{1}{2}+0-\tfrac{1}{3}+0 = -\tfrac{5}{6}$$
$$a_{13} = -\tfrac{1}{2}-\tfrac{1}{2}-\tfrac{1}{3}+0 = -\tfrac{8}{6}$$

Adding the other corresponding coefficients in the same way, we obtain the total of the four sums of squares for the four groups:

$$\begin{aligned} \phi = {} & \tfrac{13}{6}x_1^2 - \tfrac{5}{6}x_1x_2 - \tfrac{8}{6}x_1x_3 \\ & -\tfrac{5}{6}x_1x_2 + \tfrac{23}{12}x_2^2 - \tfrac{13}{12}x_2x_3 \\ & -\tfrac{8}{6}x_1x_3 - \tfrac{13}{12}x_2x_3 + \tfrac{29}{12}x_3^2 \end{aligned} \qquad (8)$$

which is another quadratic form. Thus, the sum of several quadratic forms is a quadratic form, and it has the same properties of the simple quadratic forms, namely, the diagonal square terms are positive and the off-diagonal product terms are negative. Further, the sum of the coefficients of each row (and of each column, due to symmetry) is zero. Thus, whenever we encounter a quadratic form of this nature (Chapter 4), we know it represents the total sum of squares of the three numbers

x_1, x_2, x_3 within various groups, in which any one of the xs may be duplicated or absent.

Value of a quadratic form

When the xs take certain specific numerical values, the quadratic form (8) will assume a certain value which, being a sum of squares, is always positive, except in the trivial case when $x_i = \text{constant}$. Let us calculate the value of the quadratic form (8) with $(x_1, x_2, x_3) = (-30, 24, 12)$. For numerical calculations we note that the terms of the first row of (8) have x_1 as a common factor, etc. The calculations may thus be carried out row-by-row. The row-values of (8) are:

$$\left.\begin{aligned} -30[\tfrac{13}{6}(-30) - \tfrac{5}{6}(24) - \tfrac{8}{6}(12)] &= -30(-101) - 3030 \\ 24[-\tfrac{5}{6}(-30) + \tfrac{23}{12}(24) - \tfrac{13}{12}(12)] &= \ \ 24(\ \ \ 58) = 1392 \\ 12[-\tfrac{8}{6}(-30) - \tfrac{13}{12}(24) + \tfrac{29}{12}(12)] &= \ \ 12(\ \ \ 43) = \ \ 516 \\ \text{Total } \phi &= 4938 \end{aligned}\right\} \tag{9}$$

The quadratic form (8) has the value 4938, which is the total sum of squares of the three numbers $(x_1, x_2, x_3) = (-30, 24, 12)$ within the four groups shown in (7). It is instructive to verify the result by calculating the sum of squares within each group separately, as shown in (7), and then see that they do add up to 4938 (Exercise 2).

Rank of a quadratic form

We should mention, at least briefly, the matter of rank of a quadratic form. Let a, b, c, be known coefficients. Consider the quadratic form

$$\phi = a(x_1 - x_2)^2 + b(x_1 - x_3)^2 + c(x_2 - x_3)^2 \tag{10'}$$

Although ϕ has been expressed as the sum of three squares, these three squares are not independent. For example, if $x_1 - x_2 = 9$ and $x_1 - x_3 = 13$, then $x_2 - x_3 = 4$. The lack of independence will also become obvious when we write (10′) in the fashion of a 3×3 array:

$$\begin{aligned} \phi = (a+b)x_1^2 - \quad & a x_1 x_2 - \quad & b x_1 x_3 \\ -a x_1 x_2 + \quad & (a+c)x_2^2 - \quad & c x_2 x_3 \\ -b x_1 x_3 - \quad & c x_2 x_3 + \quad & (b+c)x_3^2 \end{aligned} \tag{10}$$

When the determinant of the coefficients of a quadratic form like (10) is not zero, we say the rank of the quadratic form is 3. But in the case of (10), the determinant of the coefficients is zero, as each row is the sum of the other two rows with sign changed. In other words, the

coefficients of each row add up to zero. However, the 2×2 determinants of (10) do not vanish. Hence, we say that the rank of the quadratic form is 2.

The quadratic form (8) is of the form (10) by letting $a=\frac{5}{6}$, $b=\frac{8}{6}$, and $c=\frac{13}{12}$. Therefore, (8) is also equal to

$$\phi=\tfrac{5}{6}(x_1-x_2)^2+\tfrac{8}{6}(x_1-x_3)^2+\tfrac{13}{12}(x_2-x_3)^2 \tag{8'}$$

The reader may verify: $\phi=4938$ when $(x_1, x_2, x_3)=(-30, 24, 12)$.

When the rank of a quadratic form is r, it may be reduced to a sum of r independent squares. In our present case, the rank of (8) and (10) is 2, so it may be reduced to a sum of two independent squares. One method of accomplishing this is given in Chapter 8 when we study orthogonal contrasts among treatments.

General expressions

After some numerical practice in writing quadratic forms, the reader should be ready for limited generalizations. Let us continue to consider three numbers (x_1, x_2, x_3) for concreteness as well as for brevity. Also, suppose there are only two groups of the *x*s. Let n_{1j} be the number of times x_j appears in group I, and n_{2j} the number of times x_j appears in group II, $j=1, 2, 3$, as shown below:

	x_1	x_2	x_3	*'Size' of group*	
Group I	n_{11}	n_{12}	n_{13}	$G_1=n_{11}+n_{12}+n_{13}$	(11)
Group II	n_{21}	n_{22}	n_{23}	$G_2=n_{21}+n_{22}+n_{23}$	
	N_1	N_2	N_3		

where $N_1=n_{11}+n_{21}$ is the total number of times that x_1 appears in these two groups. We use G_1 and G_2 to denote the 'size' of the groups, that is, the number of *x*s in that group, including duplications. The notation employed here differs slightly from the conventional $n_{i\cdot}$ and $n_{\cdot j}$ but it is easier to read and write, and it provides greater distinction of the differences between the two types of totals (G_i and N_j).

First, let us write the sum of squares of the *x*s in group I in a quadratic form:

$$\begin{aligned} ssq &= n_{11}x_1^2+n_{12}x_2^2+n_{13}x_3^2-\frac{1}{G_1}(n_{11}x_1+n_{12}x_2+n_{13}x_3)^2 \\ &= n_{11}x_1^2+n_{12}x_2^2+n_{13}x_3^2 \\ &\quad -\{n_{11}^2x_1^2+n_{12}^2x_2^2+n_{13}^2x_3^2+2n_{11}n_{12}x_1x_2 \\ &\quad +2n_{11}n_{13}x_1x_3+2n_{12}n_{13}x_2x_3\}/G_1 \end{aligned}$$

In arranging these terms in a square array, we may merely indicate the coefficients of the quadratic form below, omitting the xs for brevity.

$$\begin{array}{ccc} n_{11}-\dfrac{n_{11}^2}{G_1}, & \dfrac{-n_{11}n_{12}}{G_1}, & \dfrac{-n_{11}n_{13}}{G_1} \\ \dfrac{-n_{11}n_{12}}{G_1}, & n_{12}-\dfrac{n_{12}^2}{G_1}, & \dfrac{-n_{12}n_{13}}{G_1} \\ \dfrac{-n_{11}n_{13}}{G_1}, & \dfrac{-n_{12}n_{13}}{G_1}, & n_{13}-\dfrac{n_{13}^2}{G_1} \end{array} \qquad (12)$$

These coefficients add up to zero for each row and column, as $G_1 = n_{11}+n_{12}+n_{13}$.

The sum of squares of the xs in group II is, of course, of the same form except that n_{11}, n_{12}, n_{13} of the first group are replaced by n_{21}, n_{22}, n_{23} of the second group, and G_1 is replaced by G_2. The total sum of squares for the two groups of xs is then obtained by adding the corresponding coefficients of the terms of the two quadratic forms. Using $N_1 = n_{11}+n_{21}$, etc., we see that the combined coefficients are:

$$\begin{array}{ccc} N_1-\dfrac{n_{11}^2}{G_1}-\dfrac{n_{21}^2}{G_2}, & -\dfrac{n_{11}n_{12}}{G_1}-\dfrac{n_{21}n_{22}}{G_2}, & -\dfrac{n_{11}n_{13}}{G_1}-\dfrac{n_{21}n_{23}}{G_2} \\ -\dfrac{n_{11}n_{12}}{G_1}-\dfrac{n_{21}n_{22}}{G_2}, & N_2-\dfrac{n_{12}^2}{G_1}-\dfrac{n_{22}^2}{G_2}, & \dfrac{n_{12}n_{13}}{G_1}-\dfrac{n_{22}n_{23}}{G_2} \\ -\dfrac{n_{11}n_{13}}{G_1}-\dfrac{n_{21}n_{23}}{G_2}, & -\dfrac{n_{12}n_{13}}{G_1}-\dfrac{n_{22}n_{23}}{G_2}, & N_3-\dfrac{n_{13}^2}{G_1}-\dfrac{n_{23}^2}{G_2} \end{array} \qquad (13)$$

If there is a third group of size G_3, in which x_1, x_2, x_3 appear n_{31}, n_{32}, n_{33} times, respectively, the quadratic form for the sum of squares may be written out in exactly the same way. Hence, in general, the coefficient of x_j^2 for any specified j in the combined quadratic form is

$$N_j-\frac{n_{1j}^2}{G_1}-\frac{n_{2j}^2}{G_2}-\frac{n_{3j}^2}{G_3}-\cdots = N_j-\sum_i \frac{n_{ij}^2}{G_i} \qquad (14)$$

where $N_j = n_{1j}+n_{2j}+n_{3j}+\cdots$ is the total number of times x_j appears in all the groups. Similarly, the coefficient of a product term $x_j x_{j'}$ in the square array, where $j' \neq j$, is

$$-\frac{n_{1j}n_{1j'}}{G_1}-\frac{n_{2j}n_{2j'}}{G_2}-\frac{n_{3j}n_{3j'}}{G_3}-\cdots = -\sum_i \frac{n_{ij}n_{ij'}}{G_i} \qquad (15)$$

These results will prepare the student for reading later chapters or other texts on the analysis of unbalanced data.

Exercises

1. For a group of n numbers, the sum of squares (3), when expressed in a quadratic form, takes a particularly simple pattern. The coefficient of a product term x_ix_j is always $-1/n$. The coefficient of a square term x_i^2 is always $1-1/n$, so that the sum of the coefficients in each row and column is zero. When $n=3$, the quadratic form is that given by (5). When $n=4$, the quadratic form is, writing the coefficients only,

$$\begin{pmatrix} \frac{3}{4} & -\frac{1}{4} & -\frac{1}{4} & -\frac{1}{4} \\ -\frac{1}{4} & \frac{3}{4} & -\frac{1}{4} & -\frac{1}{4} \\ -\frac{1}{4} & -\frac{1}{4} & \frac{3}{4} & -\frac{1}{4} \\ -\frac{1}{4} & -\frac{1}{4} & -\frac{1}{4} & \frac{3}{4} \end{pmatrix}$$

Now suppose that the first two numbers are the same number that appears twice, so that there are only three distinct numbers. Add the first two rows of coefficients shown above and then add the first two columns. Check that the resulting array of coefficients is the same as (6).

2. Calculate the sum of squares of each group separately, for the four groups of (7), using $(x_1, x_2, x_3)=(-30, 24, 12)$ and then see if the total sum of squares is the same as that obtained by quadratic form (8) and (9).

	Group	*Total*		*Sum of squares* (7)
I	(−30, −30, 24, 12)	−24		2520 − 144 = 2376
II	(−30, 12)	−18		1044 − 162 = 882
III	(−30, 24, 12)	6		1620 − 12 = 1608
IV	(24, 12)	36		720 − 648 = 72
			Total	5904 − 966 = 4938

The total sum of squares of the xs within the four groups is clearly

$$4x_1^2+3x_2^2+4x_3^2-\frac{(2x_1+x_2+x_3)^2}{4}-\frac{(x_1+x_3)^2}{2}$$
$$-\frac{(x_1+x_2+x_3)^2}{3}-\frac{(x_2+x_3)^2}{2}$$

which is the quadratic form (8).

3. Two groups of xs, where x_j appears n_{ij} times in group i:

	x_1	x_2	x_3	
Group I	$n_{11}=2$	$n_{12}=4$	$n_{13}=3$	$G_1=9$
Group II	$n_{21}=3$	$n_{22}=2$	$n_{23}=1$	$G_2=6$

Write out the quadratic form for the sum of squares in each group separately, and then add the two quadratic forms together.

Answer: The coefficients of the terms x_1^2, x_1x_2, etc. in the combined quadratic form are:

$$\begin{matrix} 5-\frac{4}{9}-\frac{9}{6}, & -\frac{8}{9}-\frac{6}{6}, & -\frac{6}{9}-\frac{3}{6} \\ -\frac{8}{9}-\frac{6}{6}, & 6-\frac{16}{9}-\frac{4}{6}, & -\frac{12}{9}-\frac{2}{6} \\ -\frac{6}{9}-\frac{3}{6}, & -\frac{12}{9}-\frac{2}{6}, & 4-\frac{9}{9}-\frac{1}{6} \end{matrix}$$

4. Consider two numbers x_1 and x_2 once more. Suppose that in Group I, x_1 appears a times and x_2 appears b times; and in Group II, x_1 and x_2 appear an equal number of times, $2c$, as shown below:

$$\begin{matrix} \textit{Group I} & & \textit{Group II} \\ \begin{pmatrix} x_1 & x_2 \\ a & b \end{pmatrix} & & \begin{pmatrix} x_1 & x_2 \\ 2c & 2c \end{pmatrix} \end{matrix}$$

Write out the quadratic forms for the sum of squares of each group separately, and then express it in terms of $(x_1-x_2)^2$ as we did with (4) and (4′). What is the total sum of squares within these two groups?

Note: $a-a^2/(a+b)=b-b^2/(a+b)=ab/(a+b)$.

Answer: $\phi=\left(c+\dfrac{ab}{a+b}\right)(x_1-x_2)^2$.

3

One-way classification

It is assumed that the reader has some knowledge about the analysis of variance for one-way classification data. This chapter merely reviews some of the basic calculations in order to pave the way for the analysis of two-way classification data. We shall also introduce notation that will be used or extended in subsequent chapters.

One single group (no classification)

The beginning point is to consider just one group of observations without any sub-classification. The eleven ($N=11$) numbers in the first column of Table 3.1 ($20, 31, 77, \ldots, 44$) may be thought as being the measurement of weight in grams of each of eleven mice chosen at random from a large colony of mice. Let y_α be the weight of mouse α. Then the total weight of the eleven mice is $\sum y_\alpha = Y = 550$, where the summation covers all individuals, $\alpha = 1, 2, \ldots, 11$. The mean weight of these eleven mice is defined as

$$\bar{y} = \frac{\sum y_\alpha}{N} = \frac{Y}{N} = \frac{550}{11} = 50 \qquad (1)$$

This definition makes $\sum y_\alpha = N\bar{y}$. The deviation of y_α from the mean $\bar{y}$ is $y_\alpha - \bar{y}$. The sum of the eleven deviations is $\sum (y_\alpha - \bar{y}) = \sum y_\alpha - N\bar{y} = 0$. The sum of these eleven squared deviations is $\sum (y_\alpha - \bar{y})^2 = 7338$. This number is known as the 'sum of squares' (*ssq*) of the deviations from the mean. These calculations are shown in the first four columns of Table 3.1.

In practical computation, however, the sum of squares of deviations from the mean is seldom calculated in the way described above. Rather,

it is obtained by using the algebraic identity

$$\sum (y_\alpha - \bar{y})^2 = \sum y_\alpha^2 - Y^2/N = A - C \qquad (2)$$

where

$$A = \sum y_\alpha^2 = 34\,838 \quad \text{and} \quad C = N\bar{y}^2 = Y^2/N = 27\,500 \qquad (3)$$

These calculations are shown in the two right-hand columns of Table 3.1 and we see that $A - C = 34\,838 - 27\,500 = 7338$, in agreement with the result obtained by direct calculation. In our notation, a capital letter denotes the sum of squares of the numbers themselves, for example $A = \sum y_\alpha^2$. The difference between two capital letters is a certain sum of squares required for the analysis of variance. More will be said about the notation in a later section (summary of calculations).

Model and estimation for a single group

For the data on mouse weight (first column of Table 3.1) the statistical model and its corresponding sample estimate are:

$$\text{linear model: } y_\alpha = \mu + \varepsilon_\alpha \qquad (4^*)$$

$$\text{sample estimate: } y_\alpha = m + e_\alpha \qquad (4)$$

where μ is the true (unknown) mean of the mouse population and m

Table 3.1. *Sum of squares of a single group (no classification)*

y_α	$\bar{y}$	$y_\alpha - \bar{y}$	$(y_\alpha - \bar{y})^2$	y_α^2	$\bar{y}^2$
20	50	−30	900	400	2500
31	50	−19	361	961	2500
77	50	27	729	5929	2500
72	50	22	484	5184	2500
7	50	−43	1849	49	2500
39	50	−11	121	1521	2500
28	50	−22	484	784	2500
83	50	33	1089	6889	2500
81	50	31	961	6561	2500
68	50	18	324	4624	2500
44	50	−6	36	1936	2500
550	550	0	7338	34 838	27 500
Y	$N\bar{y}$	$\sum (y_\alpha - \bar{y})$	$\sum (y_\alpha - \bar{y})^2 =$	A −	C

is an estimate of μ based on the sample observations. Once m is obtained, the residual value $e_\alpha = y_\alpha - m$ is taken as a random 'error' for the observation on mouse α.

The estimate m is to be found by the *method of least squares.* This method requires that the value of m should be so chosen that the sum of squares of the residuals e_α be a minimum. That is, the following quantity is to be minimized:

$$Q = \sum e_\alpha^2 = \sum (y_\alpha - m)^2$$

Setting

$$\frac{dQ}{dm} = -2\sum (y_\alpha - m) = 0; \quad \text{that is, } \sum (y_\alpha - m) = 0 \tag{5}$$

we obtain

$$\sum y_\alpha = Y = Nm \tag{6}$$

which is known as the *normal equation.* Solving (6),

$$m = Y/N = \bar{y} \tag{7}$$

In this simple case, the estimate m is equal to the mean $\bar{y}$. In more complicated cases, the estimate and the observed mean are not necessarily the same number; that is why we need both letters, m and $\bar{y}$. Since $m = \bar{y}$ in this case, we know that $\sum (y_\alpha - \bar{y})^2$ is the smallest among all other sums of squares $\sum (y_\alpha - a)^2$, where a is any number other than $\bar{y}$.

For a single group of observations, we estimate the population variance $\sigma^2 = E(\varepsilon^2)$ by

$$s^2 = \frac{ssq}{df} = \frac{\sum (y_\alpha - \bar{y})^2}{N-1} \tag{8}$$

where df is the number of degrees of freedom associated with the sum of squares.

We shall not develop the theoretical aspects of the analysis. In the above, our intention is to introduce the notions of the method of least squares and the resulting normal equations, and to justify the calculations of Table 3.1. In subsequent chapters the normal equations will be given without formal derivations, as these normal equations always take the form:

certain observed total = sum of corresponding parameters of the model

Equation (6) is the simplest example of a normal equation.

Subdivision of *ssq* by grouping

Suppose the large colony of mice is not a homogeneous group of one breed, but consists of four different breeds (I, II, III, IV), from which the eleven mice have been chosen: $n_1 = 4$ mice from breed I, $n_2 = 2$ mice from breed II, etc., as shown below:

I	II	III	IV
20	7	28	68
31	39	83	44
77		81	
72			

(9)

	I	II	III	IV
Group total, Y_i:	200	46	192	112
Group size, n_i:	4	2	3	2
Group mean, $\bar{y}_i$:	50	23	64	56

(10)

The eleven numbers in (9) are the same eleven numbers in the first column of Table 3.1, but have been subdivided into $k = 4$ groups. Such a division of data is called a *one-way classification*. Whenever a group of numbers is subdivided into various groups, the sum of squares of the original group may then be subdivided into two major components: ssq_W = sum of squares *within* the groups and ssq_B = sum of squares *between* the groups (with ssq_T being the total sum of squares). This we proceed to show.

Let $y_{i\alpha}$ be an observed value of mouse α in breed i, so that $y_{1\alpha}$ are the four observations in breed I, $y_{2\alpha}$ are the two observations in breed II, etc. Let us investigate the breeds separately, one at a time. Each group has its own total (Y_i), its own size (n_i = number of observations), and its own mean $(\bar{y}_i)$.

The sum of squares within groups is based on the deviations of an observation from its own group mean, that is, $(y_{1\alpha} - \bar{y}_1)$, $(y_{2\alpha} - \bar{y}_2)$, etc. Squaring each deviation and summing over the N observations, we obtain the sum of squares within groups (Table 3.2, left half):

$$\begin{aligned} ssq_W &= \sum_\alpha (y_{1\alpha} - \bar{y}_1)^2 + \cdots + \sum_\alpha (y_{4\alpha} - \bar{y}_4)^2 \\ &= \sum_{i\alpha} (y_{i\alpha} - \bar{y}_i)^2 = 5220 \end{aligned} \tag{11}$$

The sum of squares between groups is based on the deviation of a group mean from the general mean, that is, $(\bar{y}_i - \bar{y})$, of which there are n_i for the ith group. Squaring and summing, we obtain the sum of squares

between groups (Table 3.2, right half):

$$ssq_B = n_1(\bar{y}_1 - \bar{y})^2 + \cdots + n_4(\bar{y}_4 - \bar{y})^2 = \sum_i n_i(\bar{y}_i - \bar{y})^2 = 2118 \tag{12}$$

Comparing the results of Tables 3.1 and 3.2 for the same eleven numbers, we see:

$$ssq_T = ssq_W + ssq_B; \quad \text{that is, } 7338 = 5220 + 2118$$

This is always true due to the algebraic identity:

$$\sum_{i\alpha} (y_{i\alpha} - \bar{y})^2 = \sum_{i\alpha} (y_{i\alpha} - \bar{y}_i)^2 + \sum_i n_i(\bar{y}_i - \bar{y})^2 \tag{13}$$

Again, practical computation is based on algebraic identities rather than directly according to definition. Invoking identity (2), we see that the sum of squares within each group (i fixed) is $\sum_\alpha y_{i\alpha}^2 - Y_i^2/n_i$. Summing over the groups,

$$ssq_W = \sum_{i\alpha} y_{i\alpha}^2 - \sum_i \frac{Y_i^2}{n_i} = A - B \tag{14}$$

Table 3.2. *Sum of squares within and between groups*

	Within groups			*Between groups*		
	$y_{i\alpha}$	$\bar{y}_i$	$(y_{i\alpha} - \bar{y}_i)^2$	$\bar{y}_i$	$\bar{y}$	$(\bar{y}_i - \bar{y})^2$
I	20	50	900	50	50	0
	31	50	361	50	50	0
	77	50	729	50	50	0
	72	50	484	50	50	0
II	7	23	256	23	50	729
	39	23	256	23	50	729
III	28	64	1296	64	50	196
	83	64	361	64	50	196
	81	64	289	64	50	196
IV	68	56	144	56	50	36
	44	56	144	56	50	36
Total	550	550	5220	550	550	2118
			$\sum_{i\alpha} (y_{i\alpha} - \bar{y}_i)^2$			$\sum_i n_i(\bar{y}_i - \bar{y})^2$

where

$$A=\sum_i y_{i\alpha}^2 \quad \text{and} \quad B=\sum_i n_i\bar{y}_i^2=\sum_i Y_i^2/n_i \tag{15}$$

Then the sum of squares between groups is simply

$$ssq_B=\sum_i n_i(\bar{y}_i-\bar{y})^2=\sum n_i\bar{y}_i^2-N\bar{y}^2=B-C \tag{16}$$

where $C=N\bar{y}^2=Y^2/N$. In our numerical example, (9) and (10),

$$B=\frac{(200)^2}{4}+\frac{(46)^2}{2}+\frac{(192)^2}{3}+\frac{(112)^2}{2}=29\,618 \tag{15'}$$

Summary: The entire arithmetic for one-way classification reduces to the calculation of only three basic quantities: $A=34\,838$; $B=29\,618$; $C=27\,500$. The subdivision of the total sum of squares follows:

Between groups,	$ssq_B=B-C=2118$
Within groups,	$ssq_W=A-B=5220$
Total	$ssq_T=A-C=7338$

Model and estimation for one-way classification

For data of one-way classification as in (9), the statistical model and its corresponding sample estimate are:

linear model: $y_{i\alpha}=\mu+\beta_i+\varepsilon_{i\alpha}$ (17*)

sample estimate: $y_{i\alpha}=m+b_i+e_{i\alpha}$ (17)

where μ is a general constant, β_i is a constant for individuals in group i but varies from group to group, and $\varepsilon_{i\alpha}$ varies from individual to individual within a group. Again, the method of least squares requires the minimization of the quantity

$$Q=\sum_{i\alpha} e_{i\alpha}^2=\sum_{i\alpha}(y_{i\alpha}-m-b_i)^2$$

Suppose there are $k=4$ groups as in our example. There are five estimates to be made, four bs and m. Setting the partial derivatives of Q with respect to each b and m to zero, we obtain the five normal equations for the data of (9):

$$\left.\begin{array}{ll} Y_1=4m+4b_1=200; & m+b_1=50=\bar{y}_1 \\ Y_2=2m+2b_2=\ \ 46; & m+b_2=23=\bar{y}_2 \\ Y_3=3m+3b_3=192; & m+b_3=64=\bar{y}_3 \\ Y_4=2m+2b_4=112; & m+b_4=56=\bar{y}_4 \end{array}\right\} \tag{18}$$

$$Y=11m+4b_1+2b_2+3b_3+2b_4=550$$

The last equation being the sum of the preceding four, there are only four independent equations for five unknowns (m, b_1, b_2, b_3, b_4). From the discussion of equations in Chapter 1, it is clear there will be no unique solution for the five unknowns, unless some additional assumption is made. The usual practice is to take $m = \bar{y} = 50$, as in the case of having no classification. From the last equation of (18), it is seen that taking $m = \bar{y}$ is equivalent to assuming

$$\sum n_i b_i = 4b_1 + 2b_2 + 3b_3 + 2b_4 = 0 \tag{19}$$

which provides the fifth equation. Substituting $m = \bar{y} = 50$ in the first four equations of (18) we obtain a solution for the bs:

$$\left.\begin{aligned} b_1 &= \bar{y}_1 - \bar{y} = 50 - 50 = 0 \\ b_2 &= \bar{y}_2 - \bar{y} = 23 - 50 = -27 \\ b_3 &= \bar{y}_3 - \bar{y} = 64 - 50 = 14 \\ b_4 &= \bar{y}_4 - \bar{y} = 56 - 50 = 6 \end{aligned}\right\} \tag{19s}$$

which satisfies the condition (19). But the condition (19) is entirely arbitrary, as we do not possess any knowledge about the βs in the population, and the number of observations in each group is purely incidental.

Another set of solutions

We are still considering the normal equations of (18) but looking for another set of solutions for the bs by adding an equally arbitrary equation, say,

$$2b_1 + b_2 = 0 \tag{20}$$

to replace (19). Adding the first two equations of (18) and noting $4b_1 + 2b_2 = 0$, we have

$$Y_1 + Y_2 = 6m = 246; \ m = 41$$

Substituting $m = 41$ back in the right-hand equations of (18), we get a new set of solutions:

$$\left.\begin{aligned} b_1 &= 50 - 41 = 9 \\ b_2 &= 23 - 41 = -18 \\ b_3 &= 64 - 41 = 23 \\ b_4 &= 56 - 41 = 15 \end{aligned}\right\} \tag{20s}$$

In general, every arbitrary assumption of the type $\sum a_i b_i = 0$, of which (19) and (20) are examples, yields a solution for m and the bs.

However, let us note: since the values of the observed group means $\bar{y}_i$ do not change, if m is decreased (or increased) by a certain amount, the corresponding solutions for the bs must be increased (or decreased) by the same amount. In our example, m has been decreased by 9 (from 50 to 41). Hence, adding 9 to the solution (19s) yields the solution (20s). The various solutions for the bs, under various arbitrary assumptions of the type $\sum a_i b_i = 0$, differ only by a constant. In any case, $m + b_i = \bar{y}_i$ do not change, as they are specified by normal equations (18).

In theory of statistical estimation, regarding the situation described above, we say that the bs and m are not estimable from the data, as they are not uniquely determined by the data. But the joint values of $m + b_i$ are estimable, because they are equal to the observed group mean $\bar{y}_i$. The differences between the bs are also estimable, for $b_1 - b_2 = (m + b_1) - (m + b_2) = \bar{y}_1 - \bar{y}_2$. (A note on our language: what we mean is that the true parameters μ and β_i of the population are not estimable, as the sample estimates m and the bs are not uniquely determined by the data, etc. But, in subsequent chapters, we shall use the informal language for brevity.)

That the general mean of several groups is not estimable may sound baffling to non-statisticians. Perhaps a few words of intuitive 'explanation' would help. Consider the question: What is the average weight of apples and chestnuts? The question has no precise meaning without additional assumptions. What do we mean by the 'average' weight? – on a one-to-one basis?; or on a tree-to-tree basis?; or on a bushel-to-bushel basis?; or on a dollar-to-dollar basis? Without further assumptions we cannot answer the question just by weighing a number of apples and chestnuts.

In analyzing two-way classification data, a similar (actually a little worse) situation will be encountered. It is hoped that the brief description of the situation in one-way classification will ease our understanding of the more complicated cases.

Another form of *B*

The bulk of the arithmetic labor in analysis of variance in general consists of the calculation of the various sums of squares, depending on the classification pattern of the observed data. One-way classification data need the three quantities *A*, *B*, *C*, defined in (3) and (15). Of these three quantities, only *B* depends on the particular classification of the data. When the classification of the eleven numbers into four groups is done, as in (9), it has been found that $B = 29\,618$. If the same eleven

numbers were classified some other way (see Exercise 1), the values of $A = \sum y_{i\alpha}^2$ and $C = Y^2/N$ would obviously remain the same but $B = \sum n_i\bar{y}_i^2 = \sum Y_i^2/n_i$ would change, depending on the new group sizes and new group totals (or means).

From the discussion above it is clear that B is the critical value, by which the subdivision of the total sum of squares is accomplished. For one-way classification the calculation of B is simple and there is no need for further amplification. In a two-way classification situation, however, we need an additional quantity analogous to but more complicated than B; it is for this reason that we shall give another way of calculating B to pave the way for more complicated cases.

Since $m + b_i = \bar{y}_i$, although m and b_i are not known separately, we may write

$$\begin{aligned} B &= \sum \bar{y}_i Y_i = \sum (m + b_i) Y_i \\ &= mY + \sum b_i Y_i \end{aligned} \tag{21}$$

The last expression in (21) shows that B may be calculated from any particular set of solutions for m and b_i under any assumption of the type $\sum a_i b_i = 0$. In other words, the value of B is independent of the particular solutions for m and b_i. Table 3.3 gives a numerical illustration of this situation, using two different sets of (m, b_i). The analogous quantity needed in the analysis of two-way classification may be more conveniently calculated as a sum of products of type (21), which has the advantage of being independent of the particular solutions for the estimates involved.

Table 3.3. *Calculation of B as sum of products*

Estimates (19*s*)	*Estimates* (20*s*)	*Observed totals*	*Products* (19*s*) × *total*	(20*s*) × *total*
$m = 50$	$m = 41$	$Y = 550$	27 500	22 550
$b_1 = 0$	$b_1 = 9$	$Y_1 = 200$	0	1800
$b_2 = -27$	$b_2 = -18$	$Y_2 = 46$	−1242	−828
$b_3 = 14$	$b_3 = 23$	$Y_3 = 192$	2688	4416
$b_4 = 6$	$b_4 = 15$	$Y_4 = 112$	672	1680
			29 618 = B	= 29 618

Summary of calculations

The calculation of the various sums of squares presented so far has been summarized in Table 3.4. At the top of each small table is indicated what the numbers of that table are. At the bottom of each small table is indicated what the sum of squares of the numbers of that table are.

The reader will notice there are new indicators associated with letters A, B, C, in Table 3.4. Conventionally, $R(m)$ denotes the sum of squares reduced by fitting m to the data, or the sum of squares 'due to' the mean. Similarly, $R(m, b)$ denotes the sum of squares reduced by fitting m and b_i to the data, or the sum of squares due to the mean and the group effects. In this sense, we cannot write $R(m, b, e)$, which includes the error sum of squares, as we do not 'fit' the errors of the model. Instead of modifying the established meaning of the notation $R(..)$, we shall use the following informal indicators in brackets:

$$\left.\begin{aligned} A = A(m, b, e) &= \sum_{i\alpha} y_{i\alpha}^2 &&= \sum_{i\alpha} (m + b_i + e_{i\alpha})^2 \\ B = B(m, b) &= \sum_i Y_i^2/n_i &&= \sum_i n_i(m + b_i)^2 \\ C = C(m) &= Y^2/N &&= \sum m^2 \end{aligned}\right\} \qquad (22)$$

The indicators in brackets will help us to identify the meaning of a sum of squares used in the analysis of variance. Thus, $A - B$ is the sum of squares of error (that is, within groups), as $(m, b, e) - (m, b) = (e)$. Similarly, $B - C$ is the sum of squares between blocks, as $(m, b) - (m) = (b)$. For merely three quantities (A, B, C), the indicators are superfluous. But, later on, we shall have four or five quantities of such nature and the indicators will tell us the meaning of the difference between any two of them.

Analysis of variance: one-way classification

It is assumed that most readers are already familiar with the arithmetic procedure of the analysis of variance for one-way classification data. The procedure is given in Table 3.5 to complete the description of one-way classification, not as a detailed statistical explanation of the variance-ratio test.

We shall not carry out the numerical calculations with data set (9) as we shall use them again in the next chapter for two-way analysis. Hence, Table 3.5 is given in general algebriac form, where k is the number of groups.

Table 3.4. *Calculation of sum of squares for one-way classification*

$y_{i\alpha} = m + b_i + e_{i\alpha}$

20	7	28	68
31	39	83	44
77		81	
72			

$A(m, b, e) = 34\,838$

$\bar{y}_i = m + b_i$

50	23	64	56
50	23	64	56
50		64	
50			

$B(m, b) = 29\,618$

$\bar{y} = m$

50	50	50	50
50	50	50	50
50		50	
50			

$C(m) = 27\,500$

$y_{i\alpha} - \bar{y}_i = e_{i\alpha}$

−30	−16	−36	12
−19	16	19	−12
27		17	
22			

$A - B = 5220$

$\bar{y}_i - \bar{y} = b_i$

0	−27	14	6
0	−27	14	6
0		14	
0			

$B - C = 2118$

$y_{i\alpha} - \bar{y} = b_i + e_{i\alpha}$

−30	−43	−22	18
−19	−11	33	−6
27		31	
22			

$A - C = 7338$

The 'mean square' (msq) is defined as ssq/df, so that

$$s_b^2 = \frac{ssq_B}{df} = \frac{B-C}{k-1}; \qquad s_e^2 = \frac{ssq_W}{df} = \frac{A-B}{N-k}$$

The calculations exhibited in Table 3.5 are valid for random samples from any population up to the sum of squares, as it is, up to this point, merely algebraic. The next step, the calculation of s_e^2, requires an additional assumption. Let ssq_i be the sum of squares within group i with $df_i = n_i - 1$. Then $ssq_i/(n_i - 1)$ provides an estimate of the variance of ε in model (17). The assumption is that the ε of all groups have the same variance (homoscedasticity), so that we can obtain a pooled estimate of the variance of ε by pooling all ssq_i in the numerator and all df_i in the denominator. The pooled estimate is then

$$s_e^2 = \frac{\sum ssq_i}{\sum df_i} = \frac{ssq_W}{\sum (n_i - 1)} = \frac{A-B}{N-k}$$

where ssq_W is the sum of squares within the groups. It may be shown that s_b^2 is an independent estimate of the variance of ε, provided by the information on the variation of the group means $\bar{y}_i$, if the groups have no difference in true mean values (the null hypothesis).

The last step in Table 3.5, $F = s_b^2/s_e^2$, requires another assumption which is: the variables ε not only have the same variance in all breeds but are normally distributed. When this is true, the F-test is valid. But, fortunately, it has been shown that the F-test is highly 'robust'. That is, even if the ε are not exactly a normal variable, the F-test is not far wrong and is still useful for most practical purposes.

Table 3.5. *Analysis of variance for one-way classification with k groups*

Source of variation	*Degrees of freedom, df*	*Sum of squares, ssq*	*Mean square msq*	*Variance-ratio*
Between groups	$k-1$	$B-C$	s_b^2	$F = \frac{s_b^2}{s_e^2}$
Within groups	$N-k$	$A-B$	s_e^2	
Total	$N-1$	$A-C$		

The test of significance is to compare the observed F in Table 3.5 with that tabulated at a certain probability level (for example, $P=0.05$), with $k-1$ degrees of freedom for the numerator and $N-k$ degrees of freedom for the denominator of F. If the observed F is larger than the tabulated one, the difference between the groups will be judged significant. If not, it is non-significant.

Exercises

1. Suppose the same eleven numbers in (9) are classified into three groups as shown below:

		Z_i	r_i	$\bar{z}_i$
(1)	20, 31, 7, 28	86	4	21.5
(2)	77, 83, 68	228	3	76.0
(3)	72, 39, 81, 44	236	4	59.0
	Total	550	11	50.0

Since this is a different grouping from (9), we use different letters for group total (Z_i), group size (r_i), and group mean ($\bar{z}_i$). Calculate the sum of squares within and between groups.

Hint: The values of $A=\sum y_{i\alpha}^2$ and $C=Y^2/N$ given in (3) remain the same as before. Why? The only new quantity to be calculated is that analogous to B (15), for which we had also better use a new letter, say T.

$$T=\sum_i \frac{Z_i^2}{r_i}=\frac{86^2}{4}+\frac{228^2}{3}+\frac{236^2}{4}$$

Answer: between groups, $T-C=5601$; within groups, $A-T=1737$.

2. Given the following eleven numbers classified into four groups:

	26	2	32	62
	26	44	86	50
	80		74	
	68			
Y_i:	200	46	192	112	$Y=550$
n_i:	4	2	3	2	$N=\ 11$

Compare these numbers with those in (9). Construct a summary table similar to Table 3.4, using these numbers as a starting point.

Hint: In order to distinguish these new numbers from those of (9), we had better use a different capital letter for their sum of squares, say J:

$$J = 26^2 + 26^2 + 80^2 + \cdots + 62^2 + 50^2$$

The values of B and C remain the same as before. Why?

Partial answer: $J = 34\,556$, $B = 29\,618$, $C = 27\,500$.

$$J - B = 4938 \qquad B - C = 2118$$

$$J - C = 7056$$

4

Unbalanced two-way classification

Although the case of unbalanced two-way classification will be analyzed in much arithmetic detail, we shall only give the minimum calculation in this chapter, postponing the discussion on various points till later chapters.

Suppose that the eleven mice (which we have used, for example, in the previous chapter) were not only taken from four breeds but that they were subjected to three treatments. The two-way classification data are given in the upper half of Table 4.1, where a blank means an empty cell (no observation). The 'data' are far from being realistic, but our intention is to illustrate the procedure of analysis. The data do possess the feature of being unbalanced, that is, the cells have unequal numbers of observations. In this particular example, the number of observations in a cell may be 0, 1, or 2.

We may further suppose that the data arose this way. Our primary interest in the mice experiment is to compare the effects of the three treatments. It happens that there were four mice available from breed I; hence, two of them received treatment (1) and the remaining two received treatment (2) and (3) respectively. Three mice were available from breed III, each was assigned a different treatment. But only two mice were available from breed II and the same is true for breed IV. Each was assigned the treatment indicated in Table 4.1. All assignment of treatment to the mice was done at random. So the breeds will be regarded as 'blocks' employed to generate replications of the treatments which are our primary interest. The analysis in the following is geared to this situation. The linear model for the two-way classification data without interactions is

$$\text{model:} \quad y_{ij\alpha} = \mu + \beta_i + \tau_j + \varepsilon_{ij\alpha} \tag{1*}$$

$$\text{estimate:} \quad y_{ij\alpha} = m + b_i + t_j + e_{ij\alpha} \tag{1}$$

where μ is a general constant, β_i is the effect of block i, τ_j is the effect of treatment j, and $\varepsilon_{ij\alpha}$ is the random error associated with individual α who belongs to breed i and received treatment j; hence, $\alpha = 1, 2, \ldots, n_{ij}$, the number of observations in cell (ij). Model with interactions will be considered in Chapter 10.

The analysis proceeds in stages. This is nothing new to us. In the previous chapter on one-way classification, we have also proceeded in

Table 4.1. *Unbalanced two-way classification data*

	Blocks I	II	III	IV	*Treatment total* Z_j
Treatment (1)	20 31	7	28		86
Treatment (2)	77		83	68	228
Treatment (3)	72	39	81	44	236
Block total: Y_i	200	46	192	112	$550 = Y$
Block mean: $\bar{y}_i$	50	23	64	56	$50 = \bar{y}$

Preliminary calculation of *ssq* based on (one-way) classification by blocks only, ignoring the existence of the three treatments.

$A = \sum_{ij\alpha} y_{ij\alpha}^2 = 34\,838$; $B = \sum_i Y_i^2/n_i = 29\,618$; $C = Y^2/N = 27\,500$

Source	*Degrees of freedom*	*Sum of squares*	*To be calculated*
Between blocks, ignoring treatments	3	$B - C = 2118$	
Within blocks, treatments and error	7	$A - B = 5220$	Treatment *ssq*, $df = 2$ Error *ssq*, $df = 5$
Total	10	$A - C = 7338$	

stages, namely, first, ignoring the breeds, we have fitted a single general m only. Then, the classification by breeds is introduced and breed means are fitted. Here, we proceed in a similar way. First, *ignoring the treatments*, we analyze the data as if they were of one-way classification; that is, by blocks only. This has been done in the lower half of Table 4.1, where the letters A, B, C stand for the same quantities as in the previous chapter.

However, the meaning of the sums of squares has changed on account of the existence of the various treatments. Now, the sum of squares within the blocks, $A-B=5220$, with seven degrees of freedom, is no longer just due to random errors of the individuals within the blocks but is also due to the existence of the various treatments within the blocks. Hence, our major effort in the subsequent analysis is to subdivide the within-block $ssq=5220$ with $df=7$ into two components: one due to treatments with $df=2$ and another due to random error with $df=5$.

A word about our notation: We shall use b_i for block effects and t_j for treatment effects, whether the treatments (or blocks) happen to be printed as rows or columns in the data table. Hence, the subscript i (or j) does not necessarily indicate a row or a column.

Linear model and normal equations

Before calculating the treatment sum of squares (as part of the within-block ssq), let us go back to the basic model (1). The first step is to put every observed number in Table 4.1 in the form of model (1), as is done in Table 4.2.

The normal equations are derived from the principle of least squares. They are given in the lower half of Table 4.2. These normal equations are merely equating the various observed totals to the corresponding totals in terms of the parameters of the linear model. The es add up to zero for each row and for each column.

The equations for the block and treatment totals deserve attention. For treatment totals, the coefficients of m and t are always the same; thus, $Z_2=3m+3t_2+\cdots$. Similarly, for block totals, the coefficients of m and b are always the same. Consider the first block total as an example:

$$Y_1=4m+4b_1+2t_1+t_2+t_3$$

where $2t_1+t_2+t_3$ is the total of the ts in the first block. The corresponding block mean is

$$\bar{y}_1=m+b_1+\tfrac{1}{4}(2t_1+t_2+t_3)$$

Table 4.2. *Model for data of Table* 4.1 *and normal equations*

Treatment	Block I	Block II	Block III	Block IV	Treatment total
(1)	$m+t_1+b_1+e$ $m+t_1+b_1+e$	$m+t_1+b_2+e$	$m+t_1+b_3+e$		$Z_1 = 86$
(2)	$m+t_2+b_1+e$		$m+t_2+b_3+e$	$m+t_2+b_4+e$	$Z_2 = 228$
(3)	$m+t_3+b_1+e$	$m+t_3+b_2+e$	$m+t_3+b_3+e$	$m+t_3+b_4+e$	$Z_3 = 236$
Block total	$Y_1 = 200$	$Y_2 = 46$	$Y_3 = 192$	$Y_4 = 112$	$Y = 550$

The normal equations

Block totals:

$$Y_1 = 4m + 4b_1 + 2t_1 + t_2 + t_3 = 200;$$
$$Y_2 = 2m + 2b_2 + t_1 + t_3 = 46;$$
$$Y_3 = 3m + 3b_3 + t_1 + t_2 + t_3 = 192;$$
$$Y_4 = 2m + 2b_4 + t_2 + t_3 = 112;$$

Block means:

$$\bar{y}_1 = m + b_1 + \tfrac{1}{4}(2t_1 + t_2 + t_3) = 50$$
$$\bar{y}_2 = m + b_2 + \tfrac{1}{2}(t_1 + t_3) = 23$$
$$\bar{y}_3 = m + b_3 + \tfrac{1}{3}(t_1 + t_2 + t_3) = 64$$
$$\bar{y}_4 = m + b_4 + \tfrac{1}{2}(t_2 + t_3) = 56$$

Treatment totals and grand total:

$$Z_1 = 4m + 4t_1 + 2b_1 + b_2 + b_3 = 86$$
$$Z_2 = 3m + 3t_2 + b_1 + b_3 + b_4 = 228$$
$$Z_3 = 4m + 4t_3 + b_1 + b_2 + b_3 + b_4 = 236$$
$$Y = 11m + 4t_1 + 3t_2 + 4t_3 + 4b_1 + 2b_2 + 3b_3 + 2b_4 = 550$$

The coefficients of the ts in $\frac{1}{4}(2t_1+t_2+t_3)$ add up to unity, as it is the mean of the four ts in block I. These are the general properties of the normal equations for two-way classification.

At first glance, there are eight equations for the eight unknowns (m, t_1, t_2, t_3, b_1, b_2, b_3, b_4). However, it is easy to see that the eight equations in Table 4.2 are not independent. The last equation, based on the grand total, is simply the sum of the three equations based on treatment totals. It is also the sum of the four equations based on block totals. Since $Y=\sum Z_j=\sum Y_i$, there are only $8-2=6$ independent equations for eight unknowns. Hence, there will be no unique solutions for the unknowns.

For the time being, an easy way out is to introduce two additional linear equations in order to obtain unique solutions for the unknowns (although there are other procedures which make the two additional arbitrary equations unnecessary). Since this is a continuing analysis of a one-way classification data, we shall retain the general mean $m=Y/N=50$ as an estimate of μ, as if the data have no breed or treatment classifications. When we view the data this way, the two additional equations (restrictions or constraints) are

$$\sum n_ib_i=4b_1+2b_2+3b_3+2b_4=0 \qquad \text{(2b)}$$

$$\sum r_jt_j \;=4t_1+3t_2+4t_3 \qquad =0 \qquad \text{(2t)}$$

where n_i is the number of observations in block i and r_j is the number of replications of treatment j. Referring to the last normal equation (grand total) of Table 4.2, we see that the two restrictions (2b) and (2t) would simplify the equation to $Y=550=11m$, so that $m=50$. Then a solution for the ts and bs can be obtained and the sum of squares due to treatment may be calculated.

Solving for ts by eliminating bs

In order to calculate the sum of squares due to treatment, we have to obtain a solution for the ts from the normal equations. As explained in Chapter 1, this may be accomplished by eliminating the four bs from the normal equations in Table 4.2. Fortunately, the bs may be easily eliminated from the equations. For example, the equation involving the total of treatment (1) is

$$\begin{aligned}Z_1&=4t_1+4m+2b_1+b_2+b_3\\&=4t_1+2(m+b_1)+(m+b_2)+(m+b_3)\end{aligned}$$

In the meantime, we note from Table 4.2 that each block mean is of

the form $\bar{y}_i = (m + b_i) +$ terms involving ts. Substituting these block means in the Z_1-equation above, we obtain a new equation involving the ts only. Thus,

$$\begin{aligned} Z_1 = 4t_1 + 2\bar{y}_1 - \tfrac{2}{4}(2t_1 + t_2 + t_3) \\ + \bar{y}_2 - \tfrac{1}{2}(t_1 \quad + t_3) \\ + \bar{y}_3 - \tfrac{1}{3}(t_1 + t_2 + t_3) \end{aligned}$$

Transposing observed numbers to the left and collecting the coefficients of the ts, we obtain a new equation involving the ts only:

$$q_1 = Z_1 - 2\bar{y}_1 - \bar{y}_2 - \bar{y}_3 = \tfrac{13}{6}t_1 - \tfrac{5}{6}t_2 - \tfrac{8}{6}t_3 \qquad (3)$$

The numerical value of q_1 is the total of treatment (1) minus the block means as many times as t_1 appears in that block. Since t_1 appears twice in block I, we subtract $2\bar{y}_1$. Since t_1 does not appear in block IV, there is no substraction of $\bar{y}_4$ in calculating q_1.

In a similar way the bs in the Z_2- and Z_3-equations may be eliminated, so that there will be three t-equations with three unknowns (t_1, t_2, t_3). The details of the arithmetic are given in Table 4.3.

Some features of Table 4.3 are worth noting, for they are not only true for this particular example but generally true for all two-way classification situations. First, we note the q-values add up to zero:

$$\begin{aligned} q_1 &= Z_1 - 2\bar{y}_1 - \bar{y}_2 - \bar{y}_3 &= -101 \\ q_2 &= Z_2 - \bar{y}_1 - \bar{y}_3 - \bar{y}_4 &= 58 \\ q_3 &= Z_3 - \bar{y}_1 - \bar{y}_2 - \bar{y}_3 - \bar{y}_4 &= 43 \\ \hline \sum q_i &= Y - Y_1 - Y_2 - Y_3 - Y_4 &= 0 \end{aligned}$$

Second, for each q-expression, the coefficients of the three ts add up to zero; thus, $(\tfrac{13}{6}) - (\tfrac{5}{6}) - (\tfrac{8}{6}) = 0$. To see this, let us recall that the coefficients of the ts for each block mean $\bar{y}_i$ add up to unity. There are four replications for treatment (1), so that Z_1 has the term $4t_1$. From this we subtract $2\bar{y}_1 + \bar{y}_2 + \bar{y}_3$, so that the coefficients of the ts will add up to $4 - 2 - 1 - 1 = 0$. Further, the positive term of the q-expressions is always the term corresponding to the treatment total from which the q-value is calculated. In q_1, the coefficient of t_1 is positive; in q_2, the coefficient of t_2 is positive; and in q_3, the coefficient of t_3 is positive.

The purpose of constructing Table 4.3 is to obtain the three t-equations (the coefficients have been adjusted to a common denominator):

$$
\begin{aligned}
q_1 &= \tfrac{26}{12}t_1 - \tfrac{10}{12}t_2 - \tfrac{16}{12}t_3 = -101 \\
q_2 &= -\tfrac{10}{12}t_1 + \tfrac{23}{12}t_2 - \tfrac{13}{12}t_3 = 58 \\
q_3 &= -\tfrac{16}{12}t_1 - \tfrac{13}{12}t_2 + \tfrac{29}{12}t_3 = 43
\end{aligned} \tag{4}
$$

It may be further noted that the coefficients of the ts are symmetrical with respect to the principal diagonal (the three positive terms). This provides a further check on our calculations. Due to this symmetry, the coefficients in each column (fixed t_j) also add up to zero.

Table 4.3. *The elimination of the* bs *from the normal equations to obtain the t-equations*

$$
\begin{aligned}
Z_1 &= 4m + 2b_1 + b_2 + b_3 + 4t_1 = 86 \\
-2\bar{y}_1 &= -2m - 2b_1 - t_1 - \tfrac{1}{2}t_2 - \tfrac{1}{2}t_3 = -2(50) \\
-\bar{y}_2 &= -m - b_2 - \tfrac{1}{2}t_1 - \tfrac{1}{2}t_3 = -23 \\
-\bar{y}_3 &= -m - b_3 - \tfrac{1}{3}t_1 - \tfrac{1}{3}t_2 - \tfrac{1}{3}t_3 = -64
\end{aligned}
$$

$$Z_1 - 2\bar{y}_1 - \bar{y}_2 - \bar{y}_3 = q_1 = \tfrac{13}{6}t_1 - \tfrac{5}{6}t_2 - \tfrac{8}{6}t_3 = -101$$

$$
\begin{aligned}
Z_2 &= 3m + b_1 + b_3 + b_4 + 3t_2 = 228 \\
-\bar{y}_1 &= -m - b_1 - \tfrac{1}{2}t_1 - \tfrac{1}{4}t_2 - \tfrac{1}{4}t_3 = -50 \\
-\bar{y}_3 &= -m - b_3 - \tfrac{1}{3}t_1 - \tfrac{1}{3}t_2 - \tfrac{1}{3}t_3 = -64 \\
-\bar{y}_4 &= -m - b_4 - \tfrac{1}{2}t_2 - \tfrac{1}{2}t_3 = -56
\end{aligned}
$$

$$Z_2 - \bar{y}_1 - \bar{y}_3 - \bar{y}_4 = q_2 = -\tfrac{10}{12}t_1 + \tfrac{23}{12}t_2 - \tfrac{13}{12}t_3 = 58$$

$$
\begin{aligned}
Z_3 &= 4m + b_1 + b_2 + b_3 + b_4 + 4t_3 = 236 \\
-\bar{y}_1 &= -m - b_1 - \tfrac{1}{2}t_1 - \tfrac{1}{4}t_2 - \tfrac{1}{4}t_3 = -50 \\
-\bar{y}_2 &= -m - b_2 - \tfrac{1}{2}t_1 - \tfrac{1}{2}t_3 = -23 \\
-\bar{y}_3 &= -m - b_3 - \tfrac{1}{3}t_1 - \tfrac{1}{3}t_2 - \tfrac{1}{3}t_3 = -64 \\
-\bar{y}_4 &= -m - b_4 - \tfrac{1}{2}t_2 - \tfrac{1}{2}t_3 = -56
\end{aligned}
$$

$$Z_3 - \bar{y}_1 - \bar{y}_2 - \bar{y}_3 - \bar{y}_4 = q_3 = -\tfrac{16}{12}t_1 - \tfrac{13}{12}t_2 + \tfrac{29}{12}t_3 = 43$$

The equations of (4) are derived from the normal equations of Table 4.2 by eliminating m and the four bs, so that only the three ts are left. This is merely an intermediate step of the usual procedure of solving a set of linear equations.

However, when you proceed to solve for the ts from the equations of (4), you will discover that they are not three independent equations. Addition of the three equations yields $0=0$ identically. Any one of the equations may be regarded as the sum of the other two with sign reversed. There are only two independent equations for three unknowns.

In order to obtain a solution for the ts, we invoke the linear restriction equation (2t) which is independent of the equations of (4). Using any two equations (say, the first two) of (4) and (2t), we obtain three independent equations (after clearing the denominators of the coefficients):

$$\left.\begin{aligned} 13t_1 - 5t_2 - 8t_3 &= -606 \\ -10t_1 + 23t_2 - 13t_3 &= 696 \\ 4t_1 + 3t_2 + 4t_3 &= 0 \end{aligned}\right\} \qquad (5)$$

Now it is easy to solve this set of equations. In fact, we have already solved it in Chapter 1. The solution has been found to be

$$(t_1, t_2, t_3) = (-30, 24, 12) \qquad (6)$$

This particular solution depends on the particular restriction employed in the system (5). Other solutions will be presented and discussed as a separate topic in Chapter 7.

Treatment *ssq* and analysis of variance

With the q-values of (4) and the t-values of (6), the sum of squares due to treatment (as part of within-block sum of squares, usually dubbed as treatment *ssq* 'adjusted' for blocks) is readily calculated: it is simply $\sum_j q_j t_j$.

Treatments	q_j	t_j	$q_j t_j$
(1)	−101	−30	3030
(2)	58	24	1392
(3)	43	12	516
Total	0		$4938 = \sum_j q_j t_j = \phi$

(7)

In the lower portion of Table 4.1 we have already found the sum of squares within blocks: $A - B = 5220$, which is partly due to treatment

and partly due to error. Now we have found the sum of squares due to treatment in (7) and by subtraction we obtain the error sum of squares: $5220-4938=282$. The complete results are given in Table 4.4, which is an extension of the preliminary calculations made in Table 4.1.

The mean square is $msq=ssq/df$ and F is the ratio of treatment msq to error msq, which is found to be 43.8 in Table 4.4. Referring to an F-table, we find that the tabulated value is $F=13.3$, with $df=2$ for the numerator and $df=5$ for the denominator at the probability $P=0.01$ level. Since the observed $F=43.8$ is greater than the tabulated $F=13.3$, our conclusion is that the treatment differences are highly significant.

In Table 4.4 we also note that no mean square has been calculated for the 'between blocks, ignoring treatments' sum of squares. This sum of squares is obviously not entirely due to block differences. Hence, no variance-ratio test is performed for the blocks in Table 4.4. More discussion on this aspect of the problem will be found in the next chapter on dual analysis.

The assertion that the quantity $\phi=\sum t_j q_j=4938$, as given in (7), is the treatment sum of squares, may need a word of explanation. From Tables

Table 4.4. *Analysis of variance for data in Table* 4.1. *df = degrees of freedom*; *ssq = sum of squares*; *msq = mean square*

Source	*df*	*ssq*	*Further subdivision*	*df*	*ssq*	*msq*	Variance ratio
Between blocks, ignoring treatments	3	2118	(no test for block effects)				
Within blocks, treatments and error	7	5220	Treatments (adjusted)	2	4938	2469.0	$F=43.8$
			Error	5	282	56.4	
Total	10	7338					

4.1 and 4.2, we see that the treatments appear in the blocks as follows:

$$\begin{array}{cccc} \text{I} & \text{II} & \text{III} & \text{IV} \\ (t_1, t_1, t_2, t_3), & (t_1, t_3), & (t_1, t_2, t_3), & (t_2, t_3) \end{array}$$

These four groups of ts, we recognize, are the same four groups of xs studied in Chapter 2: (7). The coefficients of the present t-equations given in Table 4.3 and (4) are also the same as the coefficients of the quadratic form of Chapter 2: (8). Hence,

$$\begin{aligned} \phi &= t_1 q_1 + t_2 q_2 + t_3 q_3 \\ &= t_1(\tfrac{13}{6}t_1 - \tfrac{5}{6}t_2 - \tfrac{8}{6}t_3) \\ &\quad + t_2(-\tfrac{12}{10}t_1 + \tfrac{23}{12}t_2 - \tfrac{13}{12}t_3) \\ &\quad + t_3(-\tfrac{16}{12}t_1 - \tfrac{13}{12}t_2 + \tfrac{29}{12}t_3) \end{aligned} \tag{7'}$$

is the quadratic form representing the total sum of squares of the ts within the four blocks.

As to the value of the quadratic form, the calculations in Chapter 2: (9) and in our present (7) yield identical results:

$$\phi = 4938 \quad \text{when } (x_1, x_2, x_3) = (t_1, t_2, t_3) = (-30, 24, 12)$$

The meaning of the treatment sum of squares must be clearly grasped. The conventional terminology, treatment *ssq* ‘adjusted for blocks’, means that it is the *ssq* of the ts within the blocks. Hence, this type of analysis is also known as the *intrablock* analysis. In both Table 4.1 and Table 4.4 we have explicitly indicated that the treatment *ssq* so calculated is a part of the total within-blocks sum of squares.

That the coefficients of the t-equations, derived from eliminating the bs from the normal equations, are the same coefficients of the quadratic form representing the sum of squares of the ts within groups is a general algebraic fact. See Exercise 3. It is due to this fact that the t-equations may be written out directly from the distribution pattern of the ts within the blocks, as given in the original data (Table 4.1 or Table 4.2). The arithmetic labor of doing this is approximately the same as substituting $m + b_i$ in the normal equations.

Most experimenters are primarily interested in testing the significance of treatment differences, and will usually stop their analysis after the F-test of Table 4.4. The variance of specific comparisons between the treatments, for example $\text{Var}(t_1 - t_2)$, is given in the last section of Appendix A.

Now we shall give some further arithmetic details in subsequent paragraphs to complete the exposition of the analysis.

Values of the bs and es

The normal equations have been given in Table 4.2. It is seen that those equations have not yet been solved completely. What we have done so far is to obtain a solution for the three ts by first eliminating the m and the four bs from the equations. Now, substituting $m = 50$ and $(t_1, t_2, t_3) = (-30, 24, 12)$ back in the equations, we can solve for the four bs, thus completing the solution. This can be most easily done by using the equations for the block totals. From the equations for block totals in Table 4.2, we obtain

$$\begin{array}{llllll}
 & (Y_i) \quad (m) & (t_1) & (t_2) \; (t_3) & & \\
4b_1 = 200 - 4(50) & -2(-30) & -24 - 12 & = +24; & b_1 = +6 & \\
2b_2 = 46 - 2(50) & - (-30) & - 12 & = -36; & b_2 = -18 & \\
3b_3 = 192 - 3(50) & - (-30) & -24 - 12 & = +36; & b_3 = +12 & \quad (8) \\
2b_4 = 112 - 2(50) & & -24 - 12 & = -24; & b_4 = -12 & \\
\hline
\text{Total} \quad 4b_1 + 2b_2 + 3b_3 + 2b_4 & & & = 0 & &
\end{array}$$

The last line is the restriction (2b) on the block effects which we imposed so that $Y = 11m$ and $m = 550/11 = 50$. It also provides a check on the calculations.

At this point it should be noted that the solutions of (8) for the bs here are *not* those of the bs obtained in Chapter 3, (19s) or (20s), as they are solutions of different sets of normal equations.

Since the values of $(m, t_1, t_2, t_3, b_1, b_2, b_3, b_4)$ are all known, the values of $e_{ij\alpha}$ may be calculated. The arithmetic may be organized in two stages. The first stage is to calculate the value of $m + b_i + t_j$, by constructing the following table:

	$b_1 = 6$	$b_2 = -18$	$b_3 = 12$	$b_4 = -12$	
$m + t_1 = 20$	26 26	2	32		86
$m + t_2 = 74$	80		86	62	228
$m + t_3 = 62$	68	44	74	50	236
	200	46	192	112	550

(9)

Note that the marginal totals of the values of $m+b_i+t_j$ above are exactly the same as those of the original observed values in Table 4.1, thus providing another check on the calculations. The second stage is to obtain the values of $e_{ij\alpha}$ by subtracting the values in (9) from the original $y_{ij\alpha}$ of Table 4.1:

$$e_{ij\alpha} = y_{ij\alpha} - m - b_i - t_j$$

−6 +5	+5	−4		0
−3		−3	+6	0
+4	−5	+7	−6	0
0	0	0	0	0

(10)

The sum of squares of these e-values is

$$\sum e^2 = (-6)^2 + (5)^2 + \cdots + (7)^2 + (-6)^2 = 282 \tag{11}$$

which is precisely the value obtained in Table 4.4 by subtraction $(5220 - 4938 = 282)$. The e-values add up to zero for each row as well as for each column. This explains why the sum of squares for error has five degrees of freedom, although eleven values are exhibited above. Note that only five of the eleven values are independent. The reader may circle out five of them (for example, −6, 5, 5, in the first row and −3, −3, in the second row), and he will discover that the other six are then determined because their marginal totals are zero.

Exercises

1. The normal equations may be written out directly from the two-way table of observed values, expressed in terms of the parameters in the linear model. In the following table each dot represents an

observed value. Write out the normal equations.

	Three blocks			
	I	II	III	*Treatment total*
Treatment (1)	• •	• • • •	• • •	Z_1
Treatment (2)	• • •	• •	•	Z_2
Block total	Y_1	Y_2	Y_3	Y

Answer:

$$
\begin{aligned}
Y_1 &= 5m + 5b_1 + \qquad\qquad + 2t_1 + 3t_2 \\
Y_2 &= 6m \qquad + 6b_2 \qquad + 4t_1 + 2t_2 \\
Y_3 &= 4m \qquad\qquad + 4b_3 + 3t_1 + t_2
\end{aligned}
$$

$$
\begin{aligned}
Z_1 &= 9m + 2b_1 + 4b_2 + 3b_3 + 9t_1 \\
Z_2 &= 6m + 3b_1 + 2b_2 + \ b_3 \qquad + 6t_2
\end{aligned}
$$

$$
Y = 15m + 5b_1 + 6b_2 + 4b_3 + 9t_1 + 6t_2
$$

2. One way to review the procedure of analysis as described in this chapter is to work with another set of similar data such as that shown below:

	Blocks				
	I	II	III	IV	*Treatment total*
Treatment (1)		80	71	86	237
Treatment (2)	42	75	47	84	248
Treatment (3)	10	34 23		31	98
Block total	52	212	118	201	583
Block mean	26	53	59	67	53

Use restrictions of type (2b) and (2t), so that $m = \bar{y}$. Prepare four tables similar to Tables 4.1–4.4.

Answer: Quantities A, B, C in your first table will be new but the various sums of squares will remain the same as in Table 4.1. Your final table for the analysis of variance should be the same as Table 4.4.

3. *The derivation of the t-equations by eliminating the bs from the normal equations.* Since our purpose is to show that the coefficients of the t-equations are the same as those of the corresponding quadratic form, the reader may well review the last section of Chapter 2: general expressions. To reduce the writing to a bare minimum, we use two blocks instead of four; and we shall use the same notation as in the last section of Chapter 2, namely, n_{ij} = the number of times treatment j appears in block i. Then we have,

Treatments	(1)	(2)	(3)	*Block size*	*Block total*
Block I	$n_{11}(b_1+t_1)$	$n_{12}(b_1+t_2)$	$n_{13}(b_1+t_3)$	G_1	Y_1
Block II	$n_{21}(b_2+t_1)$	$n_{22}(b_2+t_2)$	$n_{23}(b_2+t_3)$	G_2	Y_2
Number of times t appears	N_1	N_2	N_3		
Treatment total	Z_1	Z_2	Z_3		

The es are omitted in the above because they do not enter into the normal equations. The m is omitted because m and b_i are eliminated at the same time. Then the normal equations are:

Block totals: *Block means*:

$$Y_1 = G_1 b_1 + n_{11}t_1 + n_{12}t_2 + n_{13}t_3; \qquad \bar{y}_1 = b_1 + \frac{1}{G_1}\sum_j n_{1j}t_j$$

$$Y_2 = G_2 b_2 + n_{21}t_1 + n_{22}t_2 + n_{23}t_3; \qquad \bar{y}_2 = b_2 + \frac{1}{G_2}\sum_j n_{2j}t_j$$

Treatment totals:

$$Z_1 = N_1 t_1 + n_{11}b_1 + n_{21}b_2$$

$$Z_2 = N_2 t_2 + n_{12}b_1 + n_{22}b_2$$

$$Z_3 = N_3 t_3 + n_{13}b_1 + n_{23}b_2$$

Substituting b_1 and b_2, as given by the block means, into the first treatment equation,

$$Z_1 = N_1 t_1 + n_{11}\left[\bar{y}_1 - \frac{1}{G_1}(n_{11}t_1 + n_{12}t_2 + n_{13}t_3)\right]$$
$$+ n_{21}\left[\bar{y}_2 - \frac{1}{G_2}(n_{21}t_1 + n_{22}t_2 + n_{23}t_3)\right]$$

Transposing the observed values $\bar{y}_1$ and $\bar{y}_2$ to the left to yield

$$q_1 = Z_1 - n_{11}\bar{y}_1 - n_{21}\bar{y}_2$$

and collecting the coefficients of the ts on the right-hand side,

$$q_1 = \left(N_1 - \frac{n_{11}^2}{G_1} - \frac{n_{21}^2}{G_2}\right)t_1 - \left(\frac{n_{11}n_{12}}{G_1} + \frac{n_{21}n_{22}}{G_2}\right)t_2$$
$$- \left(\frac{n_{11}n_{13}}{G_1} + \frac{n_{21}n_{23}}{G_2}\right)t_3$$

The coefficients of the ts here are identical with those in the first row of (13) of Chapter 2. The reader may write out similar expressions for q_2 and q_3 and see that the coefficients of the ts are the same as those of the quadratic form representing the sum of squares of the ts within the groups. This result may be extended to any number of treatments and any number of blocks.

In Table 4.1, the block sizes are $G_i = 4, 2, 3, 2$; and the number of times t_j appears are $N_j = 4, 3, 4$. The coefficient of t_1 in the q_1-equation is

$$N_1 - \frac{n_{11}^2}{G_1} - \frac{n_{21}^2}{G_2} - \frac{n_{31}^2}{G_3} - \frac{n_{41}^2}{G_4} = 4 - \frac{2^2}{4} - \frac{1^2}{2} - \frac{1^2}{3} - \frac{0}{2} = \frac{13}{6}$$

in agreement with what we have obtained previously in the text.

5

The dual analysis

This chapter provides some further analysis of the same unbalanced two-way classification data of the last chapter. Frequent reference to the tables and results of that chapter will be made.

In Table 4.1 the classification of blocks was introduced first, ignoring treatments. Then the treatment sum of squares was calculated as a part of the within-blocks sum of squares. This is indeed the correct procedure, if the blocks are true blocks, each block consisting of homogeneous experimental units to provide replications of the treatments.

Now, suppose that the data in Table 4.1 are derived from three blocks and four treatments. That is, the treatments and blocks are interchanged. The analysis of this new classification may be called the dual analysis (of the original one).

In order to exhibit the contrast of the new classification to the original classification, we retain the label 'treatments' for the rows and 'blocks' for the columns, remembering that these treatments now play the role of blocks and vice versa in our new analysis. The new classification is given in Table 5.1. Needless to say, the general procedure will remain the same as before, but we want to compare the results of the two analyses.

Comparing Tables 5.1 and 4.1 reveals similarities as well as differences. First, we notice the difference in classifications: solid vertical lines in Table 4.1 but solid horizontal lines in Table 5.1. Since the observed values remain the same in these two tables, the values of $A=\sum y^2$ and $C=Y^2/N$ will also remain the same as before, so that the total sum of squares remains $A-C=7338$ with ten degrees of freedom.

Regard the data as of one-way classification by treatments only. The preliminary calculation of *ssq*, ignoring the existence of the various blocks

$$A = \sum y_{ij\alpha}^2 = 34\,838; \quad T = \sum Z_j^2/r_j = 33\,101; \quad C = Y^2/N = 27\,500$$

Although the total sum of squares remains the same as before, its two components, based on primary grouping, are entirely different, as shown in the lower portions of Tables 5.1 and 4.1. Instead of calculating $B = \sum Y_i^2/n_i$ based on block totals, we now calculate $T = \sum Z_j^2/r_j$ based on treatment totals, where r_i is the number of replications of treatment i. Thus,

$$T = \sum \frac{Z_j^2}{r_j} = \frac{86^2}{4} + \frac{228^2}{3} + \frac{236^2}{4} = 33\,101 \tag{1}$$

Then, $T - C$ is the *ssq* between treatment groups (ignoring blocks) and $A - T$ is the *ssq* within treatments (partly due to block differences and

Table 5.1. *Dual analysis of the data of Table 4.1*

	Blocks				*Treatment*	
	I	II	III	IV	*total* Z_j	*mean* $\bar{z}_j$
Treatment (1)	20 31	7	28		86	21.5
Treatment (2)	77		83	68	228	76.0
Treatment (3)	72	39	81	44	236	59.0
Block total: Y_i	200	46	192	112	$550 = Y$,	$50.0 = m$

	Degrees of freedom	*Sum of squares*
Between treatments, ignoring blocks	2	$T - C = 5601$
Within treatments, blocks and error	8	$A - T = 1737\left\{\begin{matrix}\cdots\\\cdots\end{matrix}\right.$
Total	10	$A - C = 7338$

partly due to random error). Note that the subdivision of the degrees of freedom in Table 5.1 is also different from that in Table 4.1.

Subsequent work is to subdivide $A-T=1737$ with $df=8$ into two components: one due to blocks adjusted for treatments with $df=3$ and one due to error with $df=5$.

Normal equations and solution for *b*s

The normal equations for the dual analysis remain the same as those given in Table 4.2, as they are derived from the method of least squares with block effects b_i and treatment effects t_j simultaneously. Instead of writing out the expressions for the block means (as we did in Table 4.2), the reader now may wish to write out the expressions for treatment means (as exercises). A treatment mean always takes the form $\bar{z}_j = m + t_j + \cdots$, where the remaining terms involve the bs whose coefficients add up to unity. All of this should sound familiar to the reader, except for the interchange of b and t.

In order to solve for the bs, we eliminate the ts and m from the normal equations first, resulting in four equations in four unknowns (b_1, b_2, b_3, b_4). The procedure is the same as that shown in Table 4.3 except that now we want to eliminate the ts from the equations (instead of eliminating the bs from the equations as we did in Table 4.3). For example, the total for block I is

$$Y_1 = 4m + 4\,b_1 + 2t_1 + t_2 + t_3$$

Then

$$p_1 = \qquad Y_1 - 2\bar{z}_1 - \bar{z}_2 - \bar{z}_3$$

will be free of the ts and involve the bs only. Such p-values are analogous to the q-values of the last chapter (details in Table 4.3). The numerical values of the four p_is are as follows using the $\bar{z}_j$ in Table 5.1.

$$\begin{aligned}
p_1 &= Y_1 - 2\bar{z}_1 - \bar{z}_2 - \bar{z}_3 = 200 - 2(21.5) - 76 - 59 = 22.0 \\
p_2 &= Y_2 - \bar{z}_1 \quad\;\; - \bar{z}_3 = \;\, 46 - 21.5 \quad\quad\;\;\, - 59 = -34.5 \\
p_3 &= Y_3 - \bar{z}_1 - \bar{z}_2 - \bar{z}_3 = 192 - 21.5 - 76 - 59 = 35.5 \\
p_4 &= Y_4 \quad\quad\;\; - \bar{z}_2 - \bar{z}_3 = 112 \quad\quad\quad\;\; - 76 - 59 = -23.0 \\
\hline
\textstyle\sum p_i &= \textstyle\sum Y_i - Z_1 - Z_2 - Z_3 = 550 - 86 - 228 - 236 = 0
\end{aligned} \qquad (2)$$

Similar to the fact $\sum q_j = 0$ in the last chapter, the fact here, $\sum p_i = 0$, provides a check on the calculations.

Each p_i, algebraically, is free of the ts and involves only the bs. It is an instructive exercise for the reader to obtain the algebraic expressions for the four ps and then solve for the four bs. These four equations are, however, not independent; and we have to employ the restriction (2b) of the last chapter ($4b_1+2b_2+3b_3+2b_4=0$) to obtain a unique solution for the bs. The algebraic expressions for the ps may be obtained systematically in the manner of Table 4.3.

In Chapter 1, on solving linear equations, it was pointed out that the solution will remain the same no matter which unknown is eliminated first. Here we have a set of normal equations plus two additional restrictive equations. Whether we choose to eliminate the bs first and solve for the ts, or to eliminate the ts first and solve for the bs, the final solution for all the unknowns (the bs and the ts) must remain the same. If the diligent reader has already obtained the algebraic expressions for the ps (answer at the end of the chapter), he will find that the solution is

$$(b_1, b_2, b_3, b_4) = (6, -18, 12, -12) \qquad (3)$$

precisely the same as (8) of the preceding chapter, thus providing an almost fool-proof check on the correctness of his equations.

Block *ssq* and analysis of variance

With the p-values of (2) and the b-values of (3), the sum of squares due to blocks (as part of within-treatment sum of squares, usually dubbed as block *ssq* 'adjusted' for treatments) is obtained by calculating $\sum p_i b_i$ as follows:

Blocks	p_i	b_i	$p_i b_i$
I	22.0	6	132
II	−34.5	−18	621
III	35.5	12	426
IV	−23.0	−12	276
Total	0		$1455 = \sum p_i b_i$

(4)

The error sum of squares is obtained by subtraction: $1737 - 1455 = 282$. The analysis of variance is given in Table 5.2 which completes the dual analysis. From an F-table we find that the tabulated $F = 12.06$ with $df = 3$ for the numerator and $df = 5$ for the denominator at the

probability $P = .01$ level, and $F = 5.41$ at the $P = .05$ level. The observed value is $F = 8.60$, which is between the two tabulated values. Our conclusion is that the significance of the block difference is on the borderline.

Table 5.2. *Analysis of variance of data in Table* 5.1

Source	*df*	*ssq*	*Further subdivision*	*df*	*ssq*	*msq*	*Variance-ratio*
Between treatments, ignoring blocks	2	5601	(no test for treatments)				
Within treatments, blocks and error	8	1737	Blocks (adjusted)	3	1455	485.0	$F = 8.60$
			Error	5	282	56.4	
Total	10	7338		8	1737		

If the 'treatments' and 'blocks' represent two factors, both being of interest to the investigator, then we must do both analyses, one testing the factor 'treatments' and one testing the factor 'blocks'.

Comparison of the two analyses

Having finished the dual analysis we are now in a position to compare the results of Table 5.2 with those of Table 4.4. Needless to say, the total *ssq* remains the same in both cases: 7338 with 10 *df*. Moreover, we notice the error *ssq* = 282 with 5 *df* also remains the same in both tables. This means the sum of the other two components must also remain the same. This is indeed always the case:

Table 4.4	*df*	*ssq*
Between blocks, ignoring treatments	3	2118
Between treatments, adjusted for blocks	2	4938
Joint	5	7056

Table 5.2	*df*	*ssq*
Between treatments, ignoring blocks	2	5601
Between blocks, adjusted for treatments	3	1455
Joint	5	7056

This total sum of squares, 7056, is due to simultaneous block and treatment effects and may be called the joint *ssq* for blocks and treatments. This sum of squares, as a matter of fact, may be calculated directly from the t_j- and b_i-values already found in the preceding chapter. The next chapter is devoted to the investigation of the joint sum of squares (for blocks *and* treatments).

Arithmetic expediency

In the last chapter the four bs were eliminated first, resulting in three equations in ts only. In the present chapter, the three ts were eliminated first, resulting in four equations in bs only. Obviously, it is easier to solve for three unknowns from three equations than to solve for four unknowns from four equations. Since the solutions for the bs and ts remain the same in either procedure, we always prefer the case with a smaller number of equations. Thus, when there are twenty-four treatments and three blocks, we choose to eliminate the ts first from the normal equations, resulting in three equations in bs only. After the bs are found, the ts may be obtained easily. Then we calculate the *ssq* due to treatments (adjusted for blocks) as $\sum q_j t_j$.

Exercises

1. From the normal equations given in Table 4.2, write out the treatment means:

$$\bar{z}_1 = m + t_1 + \tfrac{2}{4}b_1 + \tfrac{1}{4}b_2 + \tfrac{1}{4}b_3$$

$$\bar{z}_2 = m + t_2 + \tfrac{1}{3}b_1 \qquad + \tfrac{1}{3}b_3 + \tfrac{1}{3}b_4$$

$$\bar{z}_3 = m + t_3 + \tfrac{1}{4}b_1 + \tfrac{1}{4}b_2 + \tfrac{1}{4}b_3 + \tfrac{1}{4}b_4$$

What did you notice about the coefficients of the bs in each equation?

2. In order to eliminate the ts from the normal equations we subtract, from a block total, the same number of $\bar{z}_j$ as treatment j appears in that block. The resulting equations will be in bs only. Using (2), verify the following equations:

$$
\begin{array}{rlcl}
p_1 = & Y_1 - 2\bar{z}_1 - \bar{z}_2 - \bar{z}_3 & = & \frac{29}{12}b_1 - \frac{3}{4}b_2 - \frac{13}{12}b_3 - \frac{7}{12}b_4 = 22.0 \\
p_2 = & Y_2 - \bar{z}_1 \qquad\quad - \bar{z}_3 & = & -\frac{3}{4}b_1 + \frac{3}{2}b_2 - \frac{1}{2}b_2 - \frac{1}{4}b_4 = -34.5 \\
p_3 = & Y_3 - \bar{z}_1 \; - \bar{z}_2 - \bar{z}_3 & = & -\frac{13}{12}b_1 - \frac{1}{2}b_2 + \frac{26}{12}b_3 - \frac{7}{12}b_4 = 35.5 \\
p_4 = & Y_4 \qquad\quad - \bar{z}_2 - \bar{z}_3 & = & -\frac{7}{12}b_1 - \frac{1}{4}b_2 - \frac{7}{12}b_3 + \frac{17}{12}b_4 = -23.0 \\
\hline
\sum p_i = & \sum Y_i - Z_1 - Z_2 - Z_3 & = & 0 + \;\; 0 + \;\; 0 + \;\; 0 = \;\; 0
\end{array}
$$

Only three of the four equations are independent. Using the restriction

$$4b_1 + 2b_2 + 3b_3 + 2b_4 = 0$$

solve for the bs. Check your answer with (3).

6

The joint sum of squares

It is the investigation of the joint sum of squares for blocks *and* treatments that would give us an integrated understanding of the two analyses (the original and its dual) presented in the last two chapters. When the ts are found first, the bs are then easily obtained. Conversely, when the bs are found first, the ts are then easily obtained. The calculation of the joint sum of squares (for blocks and treatments) needs the values of both the ts and the bs.

In the concluding section of Chapter 4, the values of $m+b_i+t_j$ have been calculated in order to obtain the error-values. Now, we may use them to calculate the joint sum of squares for blocks and treatments. For convenience, those numbers (Chapter 4(9)) are reproduced in the left-hand box of the following:

$m+b+t$

26 26	2	32	
80		86	62
68	44	74	50

$J=34\,556$

$b+t$

−24 −24	−48	−18	
30		36	12
18	−6	24	0

$J-C=7056$

(1)

Subtracting the general mean $m=50$ from each of the numbers in the left-hand box, we obtain the numbers labelled $b+t$ in the right-hand box of (1). The joint sum of squares for blocks and treatments is the sum of squares of the numbers labelled as $(b+t)$; thus

$$\sum (b+t)^2 = 2(-24)^2+(30)^2+\cdots+(12)^2 = 7056 \qquad (2)$$

Alternatively, we may calculate directly the *ssq* of the numbers in the left-hand box:

$$J = 2(26)^2 + (80)^2 + \cdots + (50)^2 = 34\,556 \quad (3)$$

from which subtract $C = N\bar{y}^2 = 27\,500$. Then the joint sum of squares is

$$J - C = 34\,556 - 27\,500 = 7056 \quad (4)$$

Since the total *ssq* is $A - C$ and the joint *ssq* is $J - C$, it follows that the error *ssq* is

$$\begin{aligned}(A - C) - (J - C) &= A - J \\ &= 34\,838 - 34\,556 = 282 \quad (5)\end{aligned}$$

in agreement with previous results (Table 4.4 and Table 5.2).

Practical computation

The foregoing method of calculating the joint *ssq* for blocks and treatments is purely for the purpose of exposing the meaning of that sum of squares. It is

$$J - C = \sum (m + b_i + t_j)^2 - \sum m^2 = \sum (b_i + t_j)^2 \quad (4) = (2)$$

where the summation covers all the N numbers. A practical computing method of obtaining the joint *ssq* is given in Table 6.1. As soon as the ts and bs are found, we multiply t_j by the treatment total Z_j, and multiply

Table 6.1. *Calculation of the joint sum of squares for blocks and treatments*

Parameters (estimates)	*Totals (observed)*	*Products*
$b_1 = +6$	$Y_1 = 200$	1200
$b_2 = -18$	$Y_2 = 46$	−828
$b_3 = +12$	$Y_3 = 192$	2304
$b_4 = -12$	$Y_4 = 112$	−1344
$t_1 = -30$	$Z_1 = 86$	−2580
$t_2 = +24$	$Z_2 = 228$	5472
$t_3 = +12$	$Z_3 = 236$	2832
	Total $J - C =$	7056

b_i by the block total Y_i. The sum of these two sets of products is the joint *ssq* for blocks and treatments.

$$\sum (b_i + t_j)^2 = \sum Y_i b_i + \sum Z_j t_j \tag{6}$$

Note that we have employed the observed treatment and block totals for practical computation (Table 6.1) because the general constant $m = 50$ plays no role, as $m(4b_1 + 2b_2 + 3b_3 + 2b_4) = 0$ and $m(4t_1 + 3t_2 + 4t_3) = 0$.

Subdivision of the joint *ssq*

Note that the numbers in (1) above are presented as one single group of numbers without any sub-classifications. The sum of squares (of deviations from the mean of the entire group) is found to be $J - C = 7056$. In general, *any* subgrouping imposed on the numbers of the whole group will result in a subdivision of the total sum of squares into two components: one between the groups and one within the groups. This is simply an algebraic fact. For a set of experimental data, however, we wish to group the data in a meaningful way. Thus, classification by blocks (vertically) and by treatments (horizontally) of the numbers in (1) will give us, respectively,

By blocks

26, 26	2	32	
80		86	62
68	44	74	50
Y_i: 200	46	192	112

By treatments

	Z_j
26, 26, 2, 32	86
80, 86, 62	228
68, 44, 74, 50	236

(7)

Each of these groupings will yield a subdivision of the total sum of squares 7056 into two components: between and within. The relationships exhibited in Chapter 5(5) are simply the results of the two different classifications shown above. In Table 4.4, the block (vertical) classification was adopted; in Table 5.2 the treatment (horizontal) classification was adopted.

All the necessary arithmetic has already been done and no new quantities need be calculated. The subdivision of the joint sum of squares

based on the numbers $m+b_i+t_j$ of (7) may now be summarized. The basic quantities are:

$$J=34\,556; \qquad \begin{matrix} B=\sum Y_i^2/n_i=29\,618 \\ T=\sum Z_j^2/r_j=33\,101 \end{matrix} \qquad C=27\,500 \tag{8}$$

The subdivisions of the joint *ssq* by blocks and by treatments are, respectively,

By columns		*By rows*	
Between blocks	$B-C=2118$	Between treatments	$T-C=5601$
Within blocks	$J-B=4938$	Within treatments	$J-T=1455$
Joint *ssq*	$J-C=7056$	Joint *ssq*	$J-C=7056$

(9)

which is identical with the relationship noted in Chapter 5(5). Thus, we see that the so-called treatment *ssq* 'adjusted for blocks' is really the sum of squares of the numbers $m+b_i+t_j$ within the blocks, so that it does not contain any block effects.

The calculation of the joint sum of squares for blocks and treatments also leads to a slightly different procedure of analysis. When one wishes to test the significance of treatment differences, the two different procedures of subdivision of the various sums of squares may be represented by the following sketch:

$$\begin{array}{ccc} B-C & \left.\begin{array}{c} B-C \\ \\ J-B \end{array}\right\} & J-C \\ A-B & \left\{\begin{array}{c} \\ \\ A-J \end{array}\right. & A-J \\ \hline A-C & A-C & A-C \end{array} \tag{10}$$

If we calculate the value of J, then $J-B=\phi=$ treatment *ssq* adjusted for blocks. If we calculate $\phi=\sum t_j q_j$, then there is no need to calculate J any more. The final results of these two procedures are, of course, the same. One is no better than the other, as each involves about the same amount of arithmetic.

Relationship between the ts and qs

It has been explained in Chapter 4 that the treatment *ssq*, $\phi = 4938$, is the sum of squares of the ts within the blocks. We may also say that ϕ is the sum of squares of the numbers $m + b_i + t_j$ within the blocks, as the constant $m + b_i$ for each block does not contribute to the sum of squares of the ts within the blocks. Hence, in the following we shall write the ts only. The distribution of the ts among the four blocks in our example is as follows:

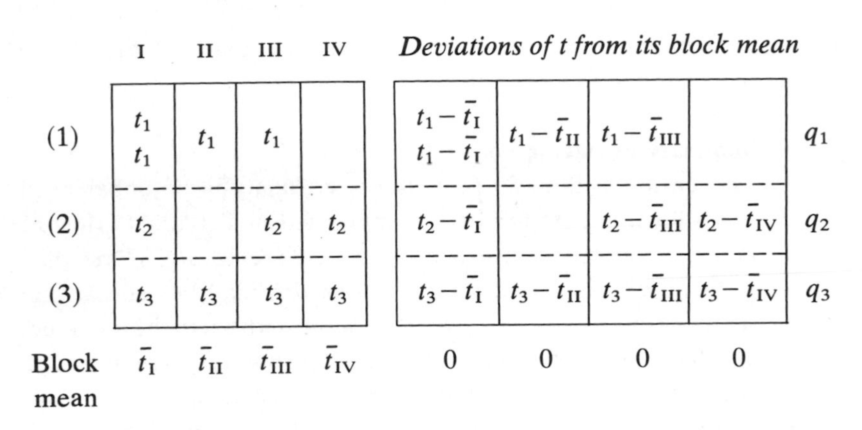

	I	II	III	IV
(1)	t_1, t_1	t_1	t_1	
(2)	t_2		t_2	t_2
(3)	t_3	t_3	t_3	t_3
Block mean	$\bar{t}_{\mathrm{I}}$	$\bar{t}_{\mathrm{II}}$	$\bar{t}_{\mathrm{III}}$	$\bar{t}_{\mathrm{IV}}$

Deviations of t from its block mean

	I	II	III	IV	
(1)	$t_1-\bar{t}_{\mathrm{I}}$, $t_1-\bar{t}_{\mathrm{I}}$	$t_1-\bar{t}_{\mathrm{II}}$	$t_1-\bar{t}_{\mathrm{III}}$		q_1
(2)	$t_2-\bar{t}_{\mathrm{I}}$		$t_2-\bar{t}_{\mathrm{III}}$	$t_2-\bar{t}_{\mathrm{IV}}$	q_2
(3)	$t_3-\bar{t}_{\mathrm{I}}$	$t_3-\bar{t}_{\mathrm{II}}$	$t_3-\bar{t}_{\mathrm{III}}$	$t_3-\bar{t}_{\mathrm{IV}}$	q_3
Block mean	0	0	0	0	

where $\bar{t}_{\mathrm{I}} = (2t_1 + t_2 + t_3)/4 =$ mean of t in block I, etc. Our previous explanation (Chapter 4) of ϕ was based on the left-hand table of ts, expressing the sum of squares within each block by a quadratic form. Now, if we take the deviation of each t from its own block mean, these deviations will add up to zero for each block, as shown in the right-hand table above. The relationship we wish to call attention to is that the row (treatment) totals of such deviations of t are precisely q_1, q_2, q_3. For example,

second row total

$$\begin{aligned} &= (t_2 - \bar{t}_{\mathrm{I}}) + (t_2 - \bar{t}_{\mathrm{III}}) + (t_2 - \bar{t}_{\mathrm{IV}}) \\ &= 3t_2 - \tfrac{1}{4}(2t_1 + t_2 + t_3) - \tfrac{1}{3}(t_1 + t_2 + t_3) - \tfrac{1}{2}(t_2 + t_3) \\ &= \tfrac{5}{6}t_1 + \tfrac{23}{12}t_2 - \tfrac{13}{12}t_3 = q_2 \end{aligned} \tag{11}$$

in agreement with Table 4.3. The arithmetic of obtaining q_j in Table 4.3 is equivalent to the present sum of deviations. That is,

$$q_2 = Z_2 - \bar{y}_1 - \bar{y}_3 - \bar{y}_4 = 3t_2 - \bar{t}_{\mathrm{I}} - \bar{t}_{\mathrm{III}} - \bar{t}_{\mathrm{IV}} \tag{12}$$

as the reader may verify by substitution.

The relationship between the t-deviations and the qs gives us a direct way of viewing the treatment ssq without resorting to quadratic forms. The treatment ssq is simply the sum of squares of the t-deviations:

$$\text{treatment } ssq = 2(t_1 - \bar{t}_{\text{I}})^2 + \cdots + (t_3 - \bar{t}_{\text{IV}})^2 \tag{13}$$

Each square may be written as $(t_1 - \bar{t}_{\text{I}})^2 = (t_1 - \bar{t}_{\text{I}})t_1 - (t_1 - \bar{t}_{\text{I}})\bar{t}_{\text{I}}$. Upon summation, the terms like $(t_1 - \bar{t}_{\text{I}})\bar{t}_{\text{I}}$ would add up to zero. The terms like $(t_1 - \bar{t}_{\text{I}})t_1$ would add up to

$$\text{treatment ssq} = q_1t_1 + q_2t_2 + q_3t_3 \tag{13'}$$

Hence, the treatment ssq is indeed the sum of squares of the eleven t-deviations from their own block means.

Summary of calculations

The entire arithmetic process of the analysis of unbalanced two-way classification data has been summarized in Table 6.2. Most of the ten tables of numbers have been obtained previously. Here, they are being put together so that their relationships may be seen at one glance. At the top of each small table is indicated what the numbers are. For instance, at the extreme left of the top row, they are indicated as $m+b+t+e$; that is, the originally observed values of y, which constitute the starting point of the calculations. At the bottom of each small table is indicated the sum of squares of the numbers in that table.

The four tables in the top row of Table 6.2 give the four basic sums of squares in descending order; the three tables in the second row are the differences between two adjacent tables. Using the 'indicators' introduced in Chapter 3, we write:

Basic quantities		*ssq = Successive differences*	
$A = A(m, b, t, e)$	$= 34\,838$	$A - J = 282$	Error
$J = J(m, b, t)$	$= 34\,556$	$J - B = 4938$	Treatments
$B = B(m, b)$	$= 29\,618$	$B - C = 2118$	Blocks
$C = C(m)$	$= 27\,500$	$A - C = 7338$	Total

The meaning of the other tables are self-evident. $A - B = 5220$ is the sum of squares for treatments and error. $J - C = 7056$ is the joint sum of squares due to blocks and treatments.

For fear of crowding, the marginal totals of the small tables in Table 6.2 are not shown. If the marginal totals of the small tables are added,

Table 6.2. *Summary of calculations for unbalanced two-way classification, assuming classification by blocks first (ignoring treatments)*

$m+b+t+e$

20 31	7	28	
77		83	68
72	39	81	44

$A = 34\,838$

$m+b+t$

26 26	2	32	
80		86	62
68	44	74	50

$J = 34\,556$

$m+b$

50 50	23	64	
50		64	56
50	23	64	56

$B = 29\,618$

m

50 50	50	50	
50		50	50
50	50	50	50

$C = 27\,500$

e

−6 5	5	−4	
−3		−3	6
4	−5	7	−6

$A-J = 282$

t

−24 −24	−21	−32		q_1
30		22	6	q_2
18	21	10	−6	q_3

$J-B = 4938$

b

0 0	−27	14	
0		14	6
0	−27	14	6

$B-C = 2118$

$t+e$

−30 −19	−16	−36		q_1
27		19	12	q_2
22	16	17	−12	q_3

$A-B = 5220$

$b+t$

−24 −24	−48	−18	
30		36	12
18	−6	24	0

$J-C = 7056$

$b+t+e$

−30 −19	−43	−22	
27		33	18
22	−11	31	−6

$A-C = 7338$

their relationships will be seen even more clearly. For example, the two left-hand tables in the top row (labelled $m+b+t+e$ and $m+b+t$, respectively) have the same row and column totals, because the values of the es add up to zero for each row and each column.

One particular table, the middle one in the second row of Table 6.2, deserves special attention. This table gives the numerical values of the t-deviations, although it has been labelled as 't'. The numbers are $t_1-\bar{t}_{\mathrm{I}}$, etc., deviations of ts from their own block means. The sum of squares of these numbers is, according to (13),

$$2(-24)^2+(30)^2+\cdots+(6)^2+(-6)^2=4938$$

which is the treatment *ssq* adjusted for blocks, correctly. The row totals of this small table are q_1, q_2, q_3.

The row and column totals of the small table labelled as '$t+e$' are the same as those of the table of 't', namely, the row totals are the qs and the column totals zero, because the es add up to zero for each row and each column.

Exercises

1. Referring to the deviations of the ts in text, show that the sum of the first row is q_1 where the qs are as given in Table 4.3 and Chapter 4(4).

Hint: Remember that there are two numbers in the first cell; thus the sum of the first row is

$$q_1=2(t_1-\bar{t}_{\mathrm{I}})+(t_1-\bar{t}_{\mathrm{II}})+(t_1-\bar{t}_{\mathrm{III}})$$

2. The situation described for the treatment *ssq* adjusted for blocks applies equally well to the block ssq adjusted for treatments, as we have done in the dual analysis (Chapter 5). Prepare a table as follows:

	I	II	III	IV	*Treatment mean*
(1)	b_1, b_1	b_2	b_3		$\bar{b}(1)$
(2)	b_1		b_3	b_4	$\bar{b}(2)$
(3)	b_1	b_2	b_3	b_4	$\bar{b}(3)$

The sum of the deviations from treatment means in the second block is

$$\begin{aligned} p_2 &= [b_2 - \bar{b}(1)] + [b_2 - \bar{b}(3)] \\ &= 2b_2 - \tfrac{1}{4}(2b_1 + b_2 + b_3) - \tfrac{1}{4}(b_1 + b_2 + b_3 + b_4) \\ &= -\tfrac{3}{4}b_1 + \tfrac{3}{2}b_2 - \tfrac{1}{2}b_3 - \tfrac{1}{4}b_4 \end{aligned}$$

in agreement with that given in Chapter 5, Exercise 2. The other p-functions may be derived in a similar way. Verify them and see that $\sum p_i b_i$ is the block sum of squares adjusted for treatments.

7

Linear restrictions; general solution

We continue to use the same set of two-way classification data in Table 4.1 for illustration. It was pointed out that only six of the eight normal equations (Table 4.2) are independent, but there are eight unknowns (m, 3ts, and 4bs). In order to obtain any numerical solutions, two additional linear equations (usually known as linear restrictions or constraints) have been introduced in Chapter 4: (2b, 2t). These two additional equations are not part of the set of normal equations derived from the principle of least squares; they are in fact quite arbitrary. Nevertheless, the calculation of the various sums of squares and the subsequent analysis of variance presented in the last three chapters are correct. The purpose of this chapter is to examine the effects of such linear constraints.

As the analysis proceeds, we soon encounter the problem of solving the equations involving the ts only (after the elimination of the bs from the normal equations). These equations (Table 4.3) are reproduced below for convenience of discussion:

$$\left.\begin{aligned} q_1 &= \tfrac{13}{6}t_1 - \tfrac{5}{6}t_2 - \tfrac{4}{3}t_3 = -101 \\ q_2 &= -\tfrac{5}{6}t_1 + \tfrac{23}{12}t_2 - \tfrac{13}{12}t_3 = 58 \\ q_3 &= -\tfrac{4}{3}t_1 - \tfrac{13}{12}t_2 + \tfrac{29}{12}t_3 = 43 \end{aligned}\right\} \tag{1}$$

As is well known now, the three equations are not independent. Two (any two) of them are, however, independent. This is where we employ a restriction in order to obtain a solution for the ts. So much for the review of our previous work.

Examples of linear constraints

The reader must have wondered by now what would happen to the analysis if we had employed some other arbitrary restrictions on the

ts. At the arithmetic level, the best way to find out is to try some other linear restrictions and compare the results with those obtained before. Let us examine a few simple ones:

$$\begin{array}{llll} \text{(i)} & 4t_1+3t_2+4t_3=0; & \text{(iv)} & t_2+t_3=0 \\ \text{(ii)} & t_1+\ t_2+\ t_3=0; & \text{(v)} & t_2 \qquad =0 \\ \text{(iii)} & 2t_1+3t_2+\ t_3=0; & \text{(vi)} & 4t_1-3t_2+2t_3=0 \end{array} \qquad (2)$$

The first restriction is of the type $\sum r_j t_j=0$, where r_j is the number of replications of treatment j. It is the one we have used in the last three chapters. The second restriction is of the type $\sum t_j=0$. It is the one almost always adopted for balanced data where $r_i=r$; that is, when all treatments have the same number of replications.

The third restriction is purely arbitrary; the coefficients are not related to r_j or any other features of the observed data. It may be regarded as an example of a general restriction of the type $\sum a_j t_j=0$, where a_j are arbitrary.

The fourth restriction, $t_2+t_3=0$, shows that a linear constraint need not involve all the ts. The fifth restriction is to set one of the ts equal to zero. This is the simplest restriction from the arithmetic point of view, for then the system (1) becomes two equations in two unknowns.

The sixth and last example in (2) deserves special attention as it involves a negative coefficient. Certain negative coefficients are permissible, while others are not. This will become clear after we examine the solutions under the various restrictions.

The solutions for the ts under the restrictions of (2) are given in the upper half of Table 7.1. Some of the solutions have already been obtained in Chapter 1, while others may be readily verified by substituting in (1). We shall give no further arithmetic.

An inspection of the six sets of solutions in the upper half of Table 7.1 reveals that they differ only by a constant. In other words, any set of solutions may be obtained from any other set by merely adding or subtracting a constant. Thus, adding 80 to the solution under (v) yields the solution under (vi), etc. Hence, once a particular set of solutions is obtained from a particular restriction, all the other solutions may be obtained; that is to say, the general solution is of the form:

$$\begin{pmatrix} t_1+c \\ t_2+c \\ t_3+c \end{pmatrix}; \quad \text{e.g.} \begin{pmatrix} -30+c \\ 24+c \\ 12+c \end{pmatrix}, \quad \begin{pmatrix} -54+c \\ 0+c \\ -12+c \end{pmatrix}, \quad \text{etc.} \qquad (3)$$

This we shall proceed to show.

General solution

Since any two of the three equations in (1) are independent, we may assign to any one of the ts an arbitrary value c and solve for the other two unknowns. Let $t_2 = c$, for example. Then the first two equations of (1) become

$$\left.\begin{aligned} \tfrac{13}{6}t_1 - \tfrac{4}{3}t_3 &= -101 + \tfrac{5}{6}c \\ -\tfrac{5}{6}t_1 - \tfrac{13}{12}t_3 &= 58 - \tfrac{23}{12}c \end{aligned}\right\} \tag{4}$$

Solving,

$$\begin{pmatrix} t_1 \\ t_2 \\ t_3 \end{pmatrix} = \begin{pmatrix} -54 + c \\ 0 + c \\ -12 + c \end{pmatrix} \tag{5}$$

which proves our claim that all solutions in the upper half of Table 7.1 differ only by a constant.

Since (3) is true, it follows that any set of particular solutions may be used to calculate the sum of squares for treatments (adjusted for blocks) which is independent of the constant on account of the fact that $\sum q_j = 0$.

$$\sum q_j(t_j + c) = \sum q_j t_j + c \sum q_j = \sum q_j t_j \tag{6}$$

This is verified in the lower half of Table 7.1. No matter which set of solutions we use, the treatment *ssq* remains 4938.

Table 7.1. *Solutions for* t*s under various restrictions and the sum of squares for treatments, adjusted for blocks*

	Solutions under restrictions (2)					
	(i)	(ii)	(iii)	(iv)	(v)	(vi)
t_1	−30	−32	−34	−48	−54	26
t_2	24	22	20	6	0	80
t_3	12	10	8	−6	−12	68
			Products $q_j t_j$			
$q_1 = -101$	3030	3232	3434	4848	5454	−2626
$q_2 = 58$	1392	1276	1160	348	0	4640
$q_3 = 43$	516	430	344	−258	−516	2924
$\sum q_j t_j$	4938	4938	4938	4938	4938	4938

Then what role does the linear restriction play? It merely specifies a particular value of c in the general solution $t_j + c$. For instance, take restriction (iv), $t_2 + t_3 = 0$. Using the general solution (5), we have

$$t_2 + t_3 = (0 + c) + (-12 + c) = 0; \quad c = 6 \tag{7}$$

and the particular solution under restriction (iv) is

$$\begin{pmatrix} t_1 \\ t_2 \\ t_3 \end{pmatrix} = \begin{pmatrix} -54+6 \\ 0+6 \\ -12+6 \end{pmatrix} = \begin{pmatrix} -48 \\ 6 \\ -6 \end{pmatrix}$$

as indicated under (iv) in Table 7.1. By this method the solution under any given restriction may be obtained directly from the general form without actually solving the equations. It is unnecessary to use the particular form $(-54 + c, c, -12 + c)$ of (5). Any particular set of ts will do, as indicated in (3).

Now we are ready to examine restriction (vi) with a negative coefficient. Using the forms of the general solution in (3), we have, according to the restriction,

$$4(-54 + c) - 3(0 + c) + 2(-12 + c) = 0; \quad c = 80 \tag{8}$$

or,

$$4(-30 + c) - 3(24 + c) + 2(12 + c) = 0; \quad c = 56 \tag{9}$$

The solution under restriction (vi) is then

$$\begin{pmatrix} t_1 \\ t_2 \\ t_3 \end{pmatrix} = \begin{pmatrix} -54+80 \\ 0+80 \\ -12+80 \end{pmatrix} = \begin{pmatrix} -30+56 \\ 24+56 \\ 12+56 \end{pmatrix} = \begin{pmatrix} 26 \\ 80 \\ 68 \end{pmatrix}$$

correctly. However, there is a limitation in using negative coefficients. Consider the restriction $2t_1 - 3t_2 + t_3 = 0$. Substituting the general solution we obtain

$$2(t_1 + c) - 3(t_2 + c) + (t_3 + c) = 0$$

which becomes the original restriction again because c is cancelled out. Such a restriction not only cannot help us to find a particular value for c but it is also inconsistent with the system (1). There will be no solutions for the ts at all.

The example above indicates the general requirement of a linear constraint with negative coefficients; that is, the coefficients of the ts should not add up to zero. Hence, equations like $t_1 - t_2 = 0$, $2t_1 - t_2 - t_3 = 0$, etc. cannot serve as linear constraints; they are not consistent with the original equations of (1). To summarize, the equation $\sum a_j t_j = 0$ can serve as a constraint provided $\sum a_j \neq 0$. Or, to put it another way, the

estimable functions (such as t_1-t_2, $2t_1-t_2-t_3$, etc.) cannot serve as restrictions; non-estimable functions (for example, $4t_1-3t_2+2t_3$) can serve as linear restrictions.

Treatment differences

From the nature of the general solution discussed above, it follows that the differences between the treatments are estimable; that is, they remain the same irrespective of the particular restriction employed. Thus, the values of

$$t_1-t_2=-54, \quad t_1-t_3=-42, \quad t_2-t_3=12 \tag{10}$$

remain the same for all solutions shown in Table 7.1.

In reviewing the sections on the sum and rank of quadratic forms of Chapter 2, we see that the treatment *ssq* adjusted for blocks may be expressed in terms of the treatment differences:

$$\begin{aligned}\phi &= a(t_1-t_2)^2+b(t_1-t_3)^2+c(t_2-t_3)^2 \\ &= \tfrac{5}{6}(-54)^2+\tfrac{4}{3}(-42)^2+\tfrac{13}{12}(12)^2=4938\end{aligned} \tag{11}$$

where a, b, c are the coefficients of the product terms t_1t_2, t_1t_3, t_2t_3 of the quadratic form ϕ.

Furthermore, since the three differences in (10) are not independent, the quadratic form for the treatment *ssq* may be expressed in terms of any two of the three differences. For instance, let $d_1=t_1-t_3$ and $d_2=t_2-t_3$. Then the quadratic form for the treatment *ssq* becomes

$$\phi=(\tfrac{13}{6}d_1-\tfrac{5}{6}d_2)d_1+(-\tfrac{5}{6}d_1+\tfrac{23}{12}d_2)d_2 \tag{12}$$

The reader should be sure that substitution of $d_1=-42$ and $d_2=12$ in quadratic form (12) does indeed yield $\phi=4938$. There are methods of reducing a quadratic form of rank r to a sum of r independent squares. This topic will be discussed in the next chapter when we study orthogonal contrasts among the treatments.

Joint values invariant

Since the treatment *ssq* remains 4938, it follows that the analysis of variance (Table 4.4) will also remain the same. We could have stopped here, as we have already shown that the linear restriction does not affect the final analysis or the test of significance of treatment differences. However, the investigation of the joint values of $m+b_i+t_j$ brings out another feature that we should be aware of. The normal equation involving the grand total of observations is

$$550=Y=11m+4t_1+3t_2+4t_3+4b_1+2b_2+3b_3+2b_4 \tag{13}$$

as given in the bottom of Table 4.2. The restrictions $\sum r_j t_j = 0$ and $\sum n_i b_j = 0$ reduce the equation to $Y = 11m$; hence, $m = Y/N = 550/11 = 50 = \bar{y}$. Under any other restrictions, the estimate m and the observed $\bar{y}$ will be two different numbers. A particular solution for the ts depends on the constraint for the ts only. A particular solution for the bs depends on the constraint for the bs only. But a solution for m depends on both constraints. In the following we investigate the consequences of restrictions of type (ii):

$$t_1 + t_2 + t_3 = 0, \quad b_1 + b_2 + b_3 + b_4 = 0 \tag{14}$$

The solutions for the ts under the condition $\sum t_i = 0$ are already known (−32, 22, 10, in Table 7.1). Substituting these values in the normal equations involving the block totals (Table 4.2), we can solve for the bs under the condition $\sum b_i = 0$.

$$\left.\begin{array}{lll} Y_1 = 4(m+b_1) - 32 = 200; & m + b_1 = 58; & b_1 = +9 \\ Y_2 = 2(m+b_2) - 22 = 46; & m + b_2 = 34; & b_2 = -15 \\ Y_3 = 3(m+b_3) + 0 = 192; & m + b_3 = 64; & b_3 = +15 \\ Y_4 = 2(m+b_4) + 32 = 112; & m + b_4 = 40; & b_4 = -9 \end{array}\right\} \tag{15}$$

$$\begin{array}{ll} 4m + 0 = 196 & \sum b_i = 0 \\ m = 49 & \end{array}$$

Alternatively, we may make use of the old solutions for the bs (6, −18, 12, −12, given in Chapter 4(8) under the original restriction $\sum n_i b_i = 0$. The general solution for the bs also takes the form $b_i + c$, where c is a constant. Then we can impose the new restriction on the general solution:

$$(6+c) + (-18+c) + (12+c) + (-12+c) = 0; \qquad c = 3 \tag{16}$$

Thus, adding 3 to the old solution yields the new solution (15). Now, we see that the three ts, four bs and m all assume new values; in particular, $m = 49 \neq \bar{y}$. The joint sum of squares for blocks and treatments (without correction for mean) is calculated in Table 7.2. It is seen that the sum of products is $J = 34\,556$, exactly the same as that obtained in Chapter 6.

That the value of J remains the same is due to the fact that the combined values of $m + b_i + t_j$ remain the same, regardless of the particular restriction employed. These values are shown below:

	$b_1=9$	$b_2=-15$	$b_3=15$	$b_4=-9$
$m+t_1=49-32=17$	26 26	2	32	
$m+t_2=49+22=71$	80		86	62
$m+t_3=49+10=59$	68	44	74	50

(17)

which is the same as in Chapter 4(9). Hence, the analysis of variance remains the same.

The fact that $m+b_i+t_j$ remains the same, whatever the linear constraints, is most easily seen from equation (13) based on the grand total of N observations. Let m be the solution for a particular set of b_i and a particular set of t_j. If a constant c is added to each b_i and a constant c' is added to each t_j in equation (13), then the new solution for m would be $m-c-c'$, so that

$$(m-c-c')+(b_i+c)+(t_j+c')=m+b_i+t_j \tag{18}$$

Table 7.2. *Calculation of the joint sum of squares for blocks and treatments (compare with Table* 6.1)

Parameters	*Totals*	*Products*
$m=49$	$Y=550$	26 950
$b_1=9$	$Y_1=200$	1800
$b_2=-15$	$Y_2=46$	−690
$b_3=15$	$Y_3=192$	2880
$b_4=-9$	$Y_4=112$	−1008
$t_1=-32$	$Z_1=86$	−2752
$t_2=22$	$Z_2=228$	5016
$t_3=10$	$Z_3=236$	2360
		$J=34\,556$

which is independent of linear restraints. Thus we say that the values of $m+b_i+t_j$ are estimable, although none of the component values is estimable separately.

A special type of restrictions

As a final example, let us consider the novel restrictions

$$\sum b_i = \sum t_j = m \tag{19}$$

Since the process of eliminating the bs from the normal equations does not make use of any restrictions, the equations of (1) remain valid whatever the restrictions. Now, with the restrictions on the ts, the equations to be solved become

$$\left.\begin{aligned} \tfrac{13}{6}t_1 - \tfrac{5}{6}t_2 - \tfrac{4}{3}t_3 &= -101 \\ -\tfrac{5}{6}t_1 + \tfrac{23}{12}t_3 - \tfrac{13}{12}t_3 &= 58 \\ t_1 + t_2 + t_3 &= m \end{aligned}\right\} \tag{20}$$

The solutions are found to be

$$\left.\begin{aligned} t_1 &= -32 + \tfrac{1}{3}m \\ t_2 &= 22 + \tfrac{1}{3}m \\ t_3 &= 10 + \tfrac{1}{3}m \end{aligned}\right\} \tag{21}$$

$$\sum t_j = 0 + m$$

Substituting these values of t into the normal equations for the block means (Table 4.2) and simplifying, we obtain

$$\left.\begin{aligned} \tfrac{4}{3}m + b_1 &= 58 \\ \tfrac{4}{3}m + b_2 &= 34 \\ \tfrac{4}{3}m + b_3 &= 64 \\ \tfrac{4}{3}m + b_4 &= 40 \end{aligned}\right\} \tag{22}$$

$$\tfrac{16}{3}m + m = 196$$

imposing the condition $\sum b_i = m$. Hence,

$$m = \tfrac{588}{19} = 30.947\,3684 \tag{23}$$

The numerical values of the bs may then be calculated from (22). However, it is desirable to put the solutions for the bs in the same form as the solutions of (21) for the ts. With $m = \frac{588}{19}$, we note

$$m + \tfrac{1}{3}m + \tfrac{1}{4}m = 49$$

which is the value of m under the restrictions $\sum b_i = \sum t_j = 0$ shown in

(15). Writing $(4/3)m = 49 - (1/4)m$ we obtain from (22):

$$\left.\begin{aligned} b_1 &= 9 + \tfrac{1}{4}m \\ b_2 &= -15 + \tfrac{1}{4}m \\ b_3 &= 15 + \tfrac{1}{4}m \\ b_4 &= -9 + \tfrac{1}{4}m \end{aligned}\right\} \tag{24}$$

$$\sum b_i = 0 + m$$

The conclusion is that the solutions under restrictions $\sum b_i = \sum t_j = m$ are the solutions under restrictions $\sum b_i = \sum t_j = 0$ plus a fraction of m as shown in (21) and (24). In Appendix A, we shall show that the solutions of the normal equations, by using the *unique* generalized inverse, always yield solutions with $\sum b_i = \sum t_j = m$.

Estimability

The investigation of estimable and non-estimable functions is beyond the scope of this book. There are methods to find or to generate estimable functions, of which there are many. However, certain simple conclusions may be drawn within the limits of the material presented in this chapter.

Obviously, the ts, the bs, and m are not individually estimable. Their values vary with the arbitrary restriction adopted. On the other hand, the treatment and block differences are estimable, because $t_1 - t_2$, $b_1 - b_2$, etc., are independent of linear restrictions, and thus uniquely determined. So are the q-functions of (1), because they are independent of the linear restrictions and assume unique numerical values.

For two-way classification data, there are two constraints: one for the ts and one for the bs. Hence, $m + b_i$ and $m + t_j$ are not estimable as their values vary with the constraints. But the joint values $m + b_i + t_j$ are estimable; they assume unique numerical values, regardless of restrictions. All linear functions of $m + b_i + t_j$ are estimable.

Further development

We have presented the analysis of two-way classification by elementary methods, without the benefit of higher mathematics, multiple regression or matrix algebra. The classical or historical approach is to adopt a convenient linear constraint for the ts, so that a solution may be obtained and the analysis of variance may be performed. Since the mid-1960s, however, the method of using a 'generalized inverse' (of a

singular matrix) to replace a linear constraint becomes prevalent. A working knowledge of matrix algebra is required to learn this comparatively new procedure. Those who have the necessary background may proceed to read Appendix A at the end of the book. We wish to establish only one thing in the appendix, namely, the correspondence between a linear restriction and a generalized inverse. For each linear restriction, and its consequent solution, we will find a generalized inverse which will yield the same solution.

Exercises

1. The basic problem in solving the normal equations is that there are eight unknowns but only six independent equations. Adding two linear restrictions to the system is equivalent to eliminating two parameters from the model. Any two parameters may be eliminated by taking their value to be zero. In some textbooks it is assumed $\mu = 0$ and $\beta_1 = 0$. In such a case we may rewrite our previous normal equations and other expressions by simply setting

$$m = 0 \quad \text{and} \quad b_1 = 0$$

The six independent normal equations are then:
Block totals:

$$Y_1 = 0 \quad + 2t_1 + t_2 + t_3 = 200; \qquad \bar{y}_1 = 50$$
$$Y_2 = 2b_2 + t_1 \quad + t_3 = 46; \qquad \bar{y}_2 = 23$$
$$Y_3 = 3b_3 + t_1 + t_2 + t_3 = 192; \qquad \bar{y}_3 = 64$$
$$Y_4 = 2b_4 \quad + t_2 + t_3 = 112; \qquad \bar{y}_4 = 56$$

Treatment totals:

$$Z_1 = 4t_1 + b_2 + b_3 \quad = 86$$
$$Z_2 = 3t_2 \quad + b_3 + b_4 = 228$$

The equation for Z_3 is not needed. Eliminating the bs from the Z-equations, the reader should obtain the following t-equations:

$$Z_1 - \bar{y}_2 - \bar{y}_3 = \tfrac{19}{6}t_1 - \tfrac{1}{3}t_2 - \tfrac{5}{6}t_3 = -1$$
$$Z_2 - \bar{y}_3 - \bar{y}_4 = \tfrac{1}{3}t_1 + \tfrac{13}{6}t_2 - \tfrac{5}{6}t_3 = 108$$

Also,

$$Y_1 = 2t_1 + t_2 + t_3 = 200$$

These three equations will yield a solution for the ts. Substituting these ts in the equations for block means, we obtain a solution for the bs.

Check your solutions with the following:

	$b_1=0$	$b_2=-24$	$b_3=6$	$b_4=-18$
$t_1=26$				empty
$t_2=80$		empty		
$t_3=68$				

Calculate the joint values b_i+t_j and check with those given in (17).

2. From the last chapter we already know $J=2(26)^2+\cdots+(50)^2=34\,556$. Verify it by calculating the sum of products of observed totals and the estimates obtained in the preceding exercise. Tabulate your arithmetic as follows:

Estimates	Observed totals	Products
$m=\ 0$	$Y=550$	
$b_1=\ 0$	$Y_1=200$	
$b_2=-24$	$Y_2=\ 46$	
$b_3=\ 6$	$Y_3=192$	
$b_4=-18$	$Y_4=112$	
$t_1=\ 26$	$Z_1=\ 86$	
$t_2=\ 80$	$Z_2=228$	
$t_3=\ 68$	$Z_3=236$	
		$J=34\,556$

3. From (23), $m=\frac{588}{19}$, $m/3=\frac{196}{19}$, $m/4=\frac{147}{19}$, and the solutions (21) and (24), show that the values of $m+b_i+t_j$ remain the same as those shown in (17). For example:

$$m+b_1+t_1=\tfrac{588}{19}+(9+\tfrac{147}{19})+(-32+\tfrac{196}{19})=26$$

Also calculate the joint sum of squares J in the manner of Table 7.2, using the solutions (21), (23), (24). See if $J=34\,556$ as before.

Hint: To avoid rounding off errors, use $19m=588$, $19b_1=318$, etc., and calculate $19J$.

8

A type of orthogonal contrasts

The concept of orthogonal contrasts is of practical importance in experimental statistics. It is a part of the general method of testing hypotheses, as orthogonal contrasts are a special type of hypotheses. A general account of hypothesis-testing procedure is beyond the level of this introductory text, but we shall give here a very simple method by which a certain type of orthogonal contrasts may be accomplished.

The orthogonal contrasts for orthogonal experimental designs, in which every treatment has the same number of replications, have been explained in most textbooks on experimental statistics. The idea may be illustrated by three treatments of effects t_1, t_2, t_3, which account for a certain sum of squares (ssq_t, say) with two degrees of freedom. Then consider the two comparisons of an orthogonal experiment:

$$2t_1 - t_2 - t_3 \quad \text{and} \quad t_2 - t_3 \tag{1}$$

The first is to compare treatment 1 versus treatments 2 and 3; the second is to compare treatment 2 versus treatment 3; each comparison will account for a certain sum of squares (ssq_1 and ssq_2) with a single degree of freedom. These two comparisons are called orthogonal contrasts because the two sums of squares due to the two comparisons of (1) will add up exactly to the sum of squares due to the three treatments; that is, $ssq_t = ssq_1 + ssq_2$, with respective df: $2 = 1 + 1$.

Further, it has been shown that the two sums of squares (ssq_1 and ssq_2), corresponding to the two comparisons of (1), are independently distributed in repeated sampling; and thus the two comparisons may be evaluated independently by the variance-ratio test. In other words, the two comparisons of (1) are two independent questions.

In unbalanced data, however, the comparisons do not take the simple form of (1) because the treatments do not appear an equal number of times in the various blocks. There are many methods of making orthogonal contrasts; most of them are complicated and require advanced knowledge in general hypothesis-testing procedures. The method to be described in this chapter is probably the simplest and follows from what we have already been familiar with.

Reduction of a quadratic form

Consider the general quadratic form in three variables that we studied in Chapter 2, reproduced here for convenience:

$$\begin{aligned}\phi = \phi(x_1, x_2, x_3) = \quad & a_{11}x_1^2 + a_{12}x_1x_2 + a_{13}x_1x_3 \\ & + a_{12}x_1x_2 + a_{22}x_2^2 + a_{23}x_2x_3 \\ & + a_{13}x_1x_3 + a_{23}x_2x_3 + a_{33}x_3^2 \end{aligned} \tag{2}$$

Now we wish to 'reduce' it into a sum of squares of certain functions of the xs, that is, into the form

$$\phi = (\cdots)^2 + (\cdots)^2 + (\cdots)^2 \tag{3}$$

where $(\cdots)$ is a function of the xs. Again, there are many methods available to accomplish this. One method is to execute the reduction in successive stages. The first step is to take, based on the first row of (2),

$$\phi_1 = \frac{1}{a_{11}}(a_{11}x_1 + a_{12}x_2 + a_{13}x_3)^2 \tag{4}$$

which takes the form $(\cdots)^2$ of (3). The coefficient $1/a_{11}$ outside the square (4) is for convenience, because we can always put $\sqrt{(1/a_{11})}$ inside the square. A similar ϕ_1 may also be written based on the second or third row of the quadratic form (2) but we shall use the first row as our example. Squaring (4) and arranging the terms according to the pattern of (2), the reader will find:

$$\begin{aligned}\phi_1 = \quad & a_{11}x_1^2 + a_{12}x_1x_2 + a_{13}x_1x_3 \\ & + a_{12}x_1x_2 + \frac{a_{12}^2}{a_{11}}x_2^2 + \frac{a_{12}a_{13}}{a_{11}}x_2x_3 \\ & + a_{13}x_1x_3 + \frac{a_{12}a_{13}}{a_{11}}x_2x_3 + \frac{a_{13}^2}{a_{11}}x_3^2 \end{aligned} \tag{4'}$$

Now note that the terms in the first row and first column of ϕ_1 are the same as those of the original ϕ. By subtraction, we would obtain a new

quadratic form involving x_2 and x_3 only. Hence,

$$\begin{aligned}\phi-\phi_1 = & \left(a_{22}-\frac{a_{12}^2}{a_{11}}\right)x_2^2 + \left(a_{23}-\frac{a_{12}a_{13}}{a_{11}}\right)x_2x_3 \\ & +\left(a_{23}-\frac{a_{12}a_{13}}{a_{11}}\right)x_2x_3 + \left(a_{33}-\frac{a_{13}^2}{a_{11}}\right)x_3^2 \qquad (5)\end{aligned}$$

With this new quadratic form (5) as a starting point, we may repeat the process to obtain another square term like $(\cdots)^2$ involving x_2 and x_3 only. Proceeding this way, we will finally accomplish the reduction (3).

The procedure of reduction described above is a general one, applicable to any quadratic form $\phi(x_1, \ldots, x_n)$. If the rank of the quadratic form is r, say, then the quadratic form may be reduced to a sum of r squares. The method of making orthogonal contrasts, to be described in the next section, is simply an application of the reduction method outlined above.

Orthogonal contrasts among treatments

The t-equations (derived from normal equations by eliminating the bs) have been given in several previous chapters. They are reproduced here, not only for convenient reference but also for generalization:

Numerical example	*In general*
$\frac{13}{6}t_1-\frac{5}{6}t_2-\frac{8}{6}t_3=-101,$	$(a+b)t_1-at_2-bt_3=q_1$
$-\frac{10}{12}t_1+\frac{23}{12}t_2-\frac{13}{12}t_3=58,$	$-at_1+(a+c)t_2-ct_3=q_2$
$-\frac{16}{12}t_1-\frac{13}{12}t_2+\frac{29}{12}t_3=43,$	$-bt_1-ct_2+(b+c)t_3=q_3$

(6)

The coefficients of each equation add up to zero. The treatment sum of squares within the blocks, in the analysis of unbalanced data, have been given as the quadratic form

$$\begin{aligned}\phi=\sum_j t_jq_j = & (a+b)t_1^2-at_1t_2-bt_1t_3 \\ & -at_1t_2-(a+c)t_2^2-ct_2t_3 \\ & -bt_1t_3-ct_2t_3-(b+c)t_3^2 \qquad (7)\end{aligned}$$

which is of the general form (2). We now want to reduce this quadratic form into a sum of squares like $(\cdots)^2+(\cdots)^2$, each square having a single degree of freedom. The rank of the quadratic form (7) is $r=2$, so it can be reduced to the sum of two squares. Applying the method (4), we have

$$\phi_1=\frac{1}{a+b}[(a+b)t_1-at_2-bt_3]^2=\frac{q_1^2}{a+b} \qquad (8)$$

where the coefficients a, b, c, are all functions of n_{ij} (last Exercise of Chapter 4). This square (or quadratic form) is due to the comparison $(a+b)t_1 - at_2 - bt_3$ which is an analogue of the comparison $2t_1 - t_2 - t_3$ for balanced data $(a = b = c)$.

If we write (8) in the pattern of a quadratic form, we will see that the coefficients of the terms of the first row and first column are the same as those of the original ϕ in (7). Thus, by subtraction, we will obtain a new quadratic form involving t_2 and t_3 only. Let $\phi_2 = \phi - \phi_1$. Then

$$\begin{aligned}\phi_2 = \; & \left(a + c - \frac{a^2}{a+b}\right)t_2^2 \; - \left(c + \frac{ab}{a+b}\right)t_2 t_3 \\ & - \left(c + \frac{ab}{a+b}\right)t_2 t_3 + \left(b + c - \frac{b^2}{a+b}\right)t_3^2 \end{aligned} \qquad (9)$$

The four coefficients in (9) are equal in absolute value:

$$a + c - \frac{a^2}{a+b} = b + c - \frac{b^2}{a+b} = c + \frac{ab}{a+b} = k$$

Hence,

$$\phi_2 = k(t_2^2 - 2t_2 t_3 + t_3^2) = \left(c + \frac{ab}{a+b}\right)(t_2 - t_3)^2 \qquad (9')$$

which is the sum of squares due to the comparison $t_2 - t_3$. To summarize:

$$\phi = \phi_1 + \phi_2 \qquad (7) = (8) + (9)$$

That is, we have subdivided the treatment *ssq* into two components: one due to the comparison of the type $2t_1 - t_2 - t_3$ and one due to the comparison of the type $t_2 - t_3$.

Finally, it remains to be shown that the two quadratic forms ϕ_1 and ϕ_2 in (8) and (9) are distributed independently, so that each comparison may be tested for significance by the F-test. The proof that they are indeed independent is given in Appendix B. Extensions to cases involving more than three treatments have been indicated in Exercise 2 and in Appendix B.

Numerical illustration

Now we return to our numerical example given in (6), where $a + b = \frac{13}{6}$, etc. No matter which set of solutions for the ts is employed (for example, $-30, 24, 12$), we always have $t_2 - t_3 = 12$, as explained in

the last chapter. Then, (8) and (9) yield

$$\phi_1 = \frac{1}{a+b} q_1^2 \qquad = \tfrac{6}{13}(-101)^2 \qquad = 4708.15$$

$$\phi_2 = \left(c + \frac{ab}{a+b}\right)(t_2 - t_3)^2 = \left(\tfrac{13}{12} + \tfrac{40}{36} \times \tfrac{6}{13}\right)(12)^2 = \quad 229.85$$

$$\phi = \phi_1 + \phi_2 \qquad = 4938.00$$

We see that most of the treatment *ssq* is due to the comparison between t_1 versus t_2 & t_3, as it is obvious from the solutions for the *t*s. The values of ϕ_1 and ϕ_2 may also be regarded as mean square (*msq*), as each has only one degree of freedom. The analysis presented here is an extension of Table 4.4. The *F*-test proceeds as follows:

	Table 4.1 *df*	*ssq*	*Table* 4.4 *df*	*ssq*	*Present analysis* *msq*	*variance ratio*
			2	4938	4708.15	$F = 83.48$
Treatment	7	5220			229.85	$F = 4.08$
and error			5	282	56.40	

The estimate of the error variance remains the same: $s^2 = \frac{282}{5} = 56.40$. The first variance-ratio is $F = \phi_1/s^2$ and the second is $F = \phi_2/s^2$. The first comparison is significant; the second is not.

This chapter gives only one simple type of orthogonal contrasts among the treatments. If the treatments are of the factorial type, the method of making orthogonal contrasts is given in Chapters 12 and 13 for factors at two levels.

Exercises

1. From the *t*-equations (6) we have three choices to take as ϕ_1; and in the text we used the first *t*-equation. Similarly, we may take the second or third equation as ϕ_1, corresponding to the comparisons of the types $2t_2 - t_1 - t_3$ and $2t_3 - t_1 - t_2$ respectively, in the orthogonal case. Thus,

$$\phi_1 = \frac{1}{a+c} q_2^2; \quad \phi_2 = \left(b + \frac{ac}{a+c}\right)(t_1 - t_3)^2$$

or

$$\phi_1 = \frac{1}{b+c} q_3^2; \quad \phi_2 = \left(a + \frac{bc}{b+c}\right)(t_1 - t_2)^2$$

Calculate the numerical values of ϕ_1 and ϕ_2, in each case, using the coefficients given in (6) and verify $\phi = \phi_1 + \phi_2$.

Answer: $4938 = 1755.13 + 3182.87$
$4938 = 765.10 + 4172.90$

2. The method described for three treatments may be easily extended to the case of four treatments. Let us use the example studied in Chapter 5, Exercise 2, as an illustration, writing t for b. Then the t-equations are

$$\begin{aligned}
q_1 &= \tfrac{1}{12}(29t_1 - 9t_2 - 13t_3 - 7t_4) = 22.0 \\
q_2 &= \tfrac{1}{4}(-3t_1 + 6t_2 - 2t_3 - t_4) = -34.5 \\
q_3 &= \tfrac{1}{12}(-13t_1 - 6t_2 + 26t_3 - 7t_4) = 35.5 \\
q_4 &= \tfrac{1}{12}(-7t_1 - 3t_2 - 7t_3 + 17t_4) = -23.0
\end{aligned}$$

It has been found in Chapter 5 that a solution for the ts is

$$(t_1, t_2, t_3, t_4) = (6, -18, 12, -12)$$

and the treatment *ssq* (adjusted) is $\phi = \sum q_j t_j = 1455$. Our present task is to decompose 1455 into three compounds, each corresponding to a certain comparison among the treatments.

Any one of the four equations may be taken as the initial comparison. If we take the first equation as the initial comparison (t_1 versus all others), then the sum of squares due to that comparison is

$$\phi_1 = \tfrac{12}{29} q_1^2 = \tfrac{12}{29}(22)^2 = 200.276$$

Write out the quadratic form for ϕ_1 and subtract it from the original ϕ for treatment *ssq*; the resulting quadratic form is (writing coefficients only):

$$\begin{pmatrix} 0 & 0 & 0 & 0 \\ 0 & 147 & -97 & -50 \\ 0 & -97 & 195 & -98 \\ 0 & -50 & -98 & 148 \end{pmatrix} \frac{1}{116}$$

involving t_2, t_3, t_4 only. Now this quadratic form assumes the same form as (7), for three treatments, and therefore the method described in the text applies. If the second comparison is t_2 versus t_3 and t_4, then we calculate

$$\begin{aligned}
\phi_2 &= \tfrac{116}{147}(147t_2 - 97t_3 - 50t_4)^2 \left(\tfrac{1}{116}\right)^2 \\
&= \frac{(3210)^2}{147 \times 116} = 604.275
\end{aligned}$$

For the remaining comparison $t_3 - t_4$ we calculate, using (9′),

$$\phi_3 = \left(\frac{98}{116} + \frac{97 \times 50/116}{147}\right)(24)^2 = 650.449$$

We see that

$$\phi = \phi_1 + \phi_2 + \phi_3 = 1455, \text{ correctly.}$$

3. The general method of reducing a quadratic form into a sum of squares is due to Lagrange (see M. Bôcher, 1938, *Introduction to Higher Algebra*, p. 131, Macmillan).

9

Analysis of another example

Having studied an example in great detail in the preceding five chapters one wonders why another numerical example is needed. Well, the new numerical example (Table 9.1) involves more ($N = 15$) observations than before, and the number of observations (n_{ij}) in a cell (i, j) ranges from 1 to 4. The work on the new numerical example gives us an opportunity to have a quick review of the essential steps of analysis. However, this, by itself, is not a sufficient reason for having another example. The chief reason for considering the new example is to pave the way for the topics of the next two chapters; *namely*, the analysis of linear models with interactions and special procedure for an experiment involving two treatments only.

The linear model for the present example is the same as before:

$$y_{ij\alpha} = m + b_i + t_j + e_{ij\alpha} \tag{1}$$

where $\alpha = 1, 2, \ldots, n_{ij}$, the number of observations in a cell. The normal equations for the observed data in the upper portion of Table 9.1 are:

Block totals: *Block means*:

$$Y_1 = 5m + 5b_1 + 2t_1 + 3t_2 = 120; \quad \bar{y}_1 = m + b_1 + (2t_1 + 3t_2)/5 = 24.00$$

$$Y_2 = 6m + 6b_2 + 4t_1 + 2t_2 = 86; \quad \bar{y}_2 = m + b_2 + (4t_1 + 2t_2)/6 = 14.33$$

$$Y_3 = 4m + 4b_3 + 3t_1 + t_2 = 94; \quad \bar{y}_3 = m + b_3 + (3t_1 + t_2)/4 = 23.50$$

Treatment totals and grand total: (2)

$$Z_1 = 9m + 2b_1 + 4b_2 + 3b_3 + 9t_1 = 143$$

$$Z_2 = 6m + 3b_1 + 2b_2 + b_3 + 6t_2 = 157$$

$$Y = 15m + 5b_1 + 6b_2 + 4b_3 + 9t_1 + 6t_2 = 300$$

Table 9.1. *Preliminary calculations in the analysis of two-way classification data. One-way classification by blocks only*

	Block I	Block II	Block III	Total Z_j	Mean $\bar{z}_j$
Treatment (1)	27 15	9 17 6 18	14 22 15	143	15.88
Treatment (2)	33 16 29	10 26	43	157	26.16
Block total: Y_i	120	86	94	300	
Block mean: $\bar{y}_i$	24.00	14.33	23.50		$\bar{y} = 20$

$$A = \sum_i \sum_j \sum_\alpha y_{ij\alpha}^2 = 27^2 + 15^2 + \cdots + 43^2 = 7400.00$$

$$B = \sum_j \frac{Y_i^2}{n_i} = \frac{120^2}{5} + \frac{86^2}{6} + \frac{94^2}{4} = 6321.67$$

$$C = Y^2/N = (300)^2/15 = 6000.00$$

Source of variation	*df*	*ssq*	*To be subdivided*
Between blocks, ignoring treatments	2	$B - C = 321.67$	
Within blocks, treatments and error	12	$A - B = 1078.33$	Treatments, adjusted for blocks; Error
Total	14	$A - C = 1400.00$	

Only four of the six equations are independent. The two linear constraints we choose to use are

$$5b_1+6b_2+4b_3=0, \quad 9t_1+6t_2=0 \tag{3}$$

so that $m = Y/N = 300/15 = 20$.

The preliminary calculations are also given in Table 9.1. First, we ignore the existence of treatments and regard the data as of one-way classification (by blocks only). Hence, the calculations of the quantities A, B, C. The within-block $ssq = A - B = 1078.33$ is to be subdivided into two components: treatment ssq (adjusted for blocks) and error ssq with $df = 1$ and $df = 11$, respectively.

Treatment sum of squares

To solve for the ts, we eliminate the bs from the equations for treatment totals. Thus,

$$\left.\begin{aligned} q_1 &= Z_1-2\bar{y}_1-4\bar{y}_2-3\bar{y}_3 \\ &= 143-2(24)-4(14.33)-3(23.50) = -32.83 \\ q_2 &= Z_2-3\bar{y}_1-2\bar{y}_2-\ \bar{y}_3 \\ &= 157-3(24)-2(14.33)-\ (23.50) = +32.83 \end{aligned}\right\} \tag{4}$$

Since $q_1+q_2=0$, the calculation of q_2 is not necessary; it is given above to check the arithmetic. We only need to obtain the algebraic expression for q_1. Since the bs and m are all cancelled out, we will write out the terms involving the ts only. Thus,

$$q_1 = 9t_1 - 2\left(\frac{2t_1+3t_2}{5}\right) - 4\left(\frac{4t_1+2t_2}{6}\right) - 3\left(\frac{3t_1+t_2}{4}\right)$$

Simplifying,

$$\left.\begin{aligned} q_1 &= \frac{197}{60}t_1 - \frac{197}{60}t_2 = -32.833 \\ q_2 &= \frac{-197}{60}t_1 + \frac{197}{60}t_2 = +32.833 \end{aligned}\right\} \tag{5}$$

As $60 \times 32.83 = 1970$, both equations reduce to

$$t_1 - t_2 = -10 \tag{5'}$$

This, together with restriction (3), $3t_1+2t_2=0$, yields the solution

$$t_1 = -4, \quad t_2 = 6 \tag{5s}$$

The treatment *ssq*, adjusted for blocks, is calculated in the usual manner:

	q_j	t_j	$q_j t_j$
(1)	−32.83	−4	131.33
(2)	+32.83	+6	197.00
		Treatment $ssq = \sum q_j t_j =$	328.33

(6)

Since there is only one degree of freedom for two treatments, the treatment $ssq = \sum q_j t_j$ may also be expressed as one square; thus

$$q_1 t_1 + q_2 t_2 = \tfrac{197}{60}(t_1 - t_2)t_1 - \tfrac{197}{60}(t_1 - t_2)t_2$$
$$= \tfrac{197}{60}(t_1 - t_2)^2 = \tfrac{197}{60}(-10)^2 = 328.33 \qquad (6')$$

Joint *ssq* and error *ssq*

Substituting $m = 20$, $t_1 = -4$, $t_2 = 6$, in the normal equations for block totals of (2), we obtain

$$Y_1 = 5(20) + 5b_1 + 2(-4) + 3(6) = 120; \; 5b_1 = 10, \; b_1 = 2$$
$$Y_2 = 6(20) + 6b_2 + 4(-4) + 2(6) = 86; \; 6b_2 = -30, \; b_2 = -5 \qquad (7)$$
$$Y_3 = 4(20) + 4b_3 + 3(-4) + (6) = 94; \; 4b_3 = 20, \; b_3 = 5$$
$$\sum n_i b_i = 0$$

The values of $m + b_i + t_j$ are as follows:

$m + t$	$b_1 = 2$	$b_2 = -5$	$b_3 = 5$	
$20 - 4 = 16$	18 18	11 11 11 11	21 21 21	143
$20 + 6 = 26$	28 28 28	21 21	31	157
Block total	120	86	94	300

(8)

The sum of squares of these numbers

$$J = 2(18)^2 + 4(11)^2 + \cdots + (31)^2 = 6650 \qquad (9)$$
$$J - C = 6650 - 6000 = 650 \qquad (10)$$

is the joint sum of squares for blocks and treatments. This is in agreement with the results of Table 9.2, in which

between blocks, ignoring treatments	= 321.67
treatments, adjusted for blocks	= 328.33
joint *ssq* for blocks and treatments	= 650.00

(10′)

Table 9.2. *Analysis of variance of data in Table 9.1 without interaction*

	df	*ssq*		*df*	*ssq*	*msq*
Between blocks, ignoring treatments	2	321.67			(no test for blocks)	
Within blocks, and error	12	1078.33	Treatments adjusted for blocks	1	328.33	328.33
			Error	11	750.00	68.18
Total	14	1400.00		12	1078.33	$F = 4.82$

The residual value is $e_{ij\alpha} = y_{ij\alpha} - (m + b_i + t_j)$, where $y_{ij\alpha}$ is given in Table 9.1 and $m + b_i + t_j$ is given in (8). Thus, the e-values are:

	I	II	III	
(1)	9 −3	−2 6 −5 7	−7 1 −6	0
(2)	5 −12 1	−11 5	12	0
	0	0	0	0

(11)

The marginal totals of (8) are the same as those of the observed data in Table 9.1, so the e-values add up to zero for each row and for each column. The sum of squares of these values is

$$\sum e^2 = 9^2 + (-3)^2 + \cdots + 12^2 = 750 \tag{12}$$

which, again, is in agreement with the result obtained by subtraction in Table 9.2. All of these should be quite familiar to the reader who has studied Chapter 4. It is these e-values of (11) and their sum of squares (12) that we shall further investigate in the next chapter (model with interactions).

Special expression for two treatments

The coefficients of the ts in expression (5) for q_1 are all functions of n_{ij}, the number of observations in a cell (ij). These numbers in our example are

	I	II	III
Treatment (1)	$n_{11}=2$,	$n_{12}=4$,	$n_{13}=3$
Treatment (2)	$n_{21}=3$,	$n_{22}=2$,	$n_{23}=1$
	$n_{11}+n_{21}=5$,	$n_{12}+n_{22}=6$,	$n_{13}+n_{23}=4$

Hence, in general, after eliminating the bs and m from the normal equations, we have

$$q_1 = (n_{11}+n_{12}+n_{13})t_1 - n_{11}\left(\frac{n_{11}t_1+n_{21}t_2}{n_{11}+n_{21}}\right) - n_{12}\left(\frac{n_{12}t_1+n_{22}t_2}{n_{12}+n_{22}}\right) - n_{13}\left(\frac{n_{13}t_1+n_{23}t_2}{n_{13}+n_{23}}\right) \tag{5'}$$

Substituting the numerical values of n_{ij}, the reader will see that the general expression above reduces to (5). The coefficients of t_2, all negative, are easily seen. The coefficients of t_1 turn out to be the same as those of t_2 but opposite in sign (that is, all positive). For instance, the coefficient for t_1 of the first block is

$$n_{11} - n_{11}\left(\frac{n_{11}}{n_{11}+n_{21}}\right) = \frac{n_{11}n_{21}}{n_{11}+n_{21}}$$

which is also the coefficient of t_2 of the first block, with sign reversed. The same relationships hold for the ts of the second and third blocks.

Hence,

$$q_1 = \left(\frac{n_{11}n_{21}}{n_{11}+n_{21}} + \frac{n_{12}n_{22}}{n_{12}+n_{22}} + \frac{n_{13}n_{23}}{n_{13}+n_{23}}\right)(t_1 - t_2)$$

$$= (w_1 + w_2 + w_3)(t_1 - t_2) \qquad (5'')$$

where

$$w_j = \frac{n_{1j}n_{2j}}{n_{1j}+n_{2j}}$$

In our numerical example, $w_1 = \frac{6}{5}$, $w_2 = \frac{8}{6}$, $w_3 = \frac{3}{4}$, so that

$$\sum w_j = \tfrac{6}{5} + \tfrac{8}{6} + \tfrac{3}{4} = \tfrac{197}{60}$$

in agreement with the coefficient in (5). The treatment *ssq* (adjusted for blocks) is then

$$\text{treatment } ssq = (\sum w_j)(t_1 - t_2)^2 \qquad (6'')$$

which agrees with (6′). It will be seen that the special procedure for comparing two treatments only (Chapter 11) depends on the general expression (5″).

Exercises

1. Do the dual analysis of variance on the same set of data (Table 9.1) to test the significance of block differences.

Hint: Relax; there is practically nothing to do. The values of A, J, C all remain the same; only one new quantity (to take the place of B) is needed:

$$T = \sum_i \frac{Z_i^2}{r_i} = \frac{143^2}{9} + \frac{157^2}{6} = 6380.27$$

Answer:

	df	*ssq*		*df*	*ssq*	*msq*
Between treatments, ignoring blocks	1	$T - C = 380.27$	(no test for treatments)			
Within treatments, blocks and error	13	$A - T = 1019.72$	Blocks, adjusted for treatments	2	$J - T = 269.72$	134.86
			Error	11	$A - J = 750.00$	68.18
Total	14	$A - C = 1400.00$				$F = 1.98$

2. In the exercise above, the sum of squares due to blocks (adjusted for treatments) has been found to be 269.72, using the known value $J=6650$ given in (9). If the reader wishes, it may be independently verified, using the solution for the *b*s (7). Then we calculate (Chapter 5):

Block total (treatment mean)	p_i	b_i	$p_i b_i$
$p_1 = 120 - 2(15.88) - 3(26.16) =$	$+9.72$	2	19.44
$p_2 = 86 - 4(15.88) - 2(26.16) =$	-29.88	-5	149.44
$p_3 = 94 - 3(15.88) - (26.16) =$	$+20.16$	5	100.83
$\sum p_i = 300 - 143 \quad - 157 \quad =$	0		269.72

10

Model with interactions

This is a continuation of the analysis of the data in Table 9.1. The analysis of linear models without interactions amounts to the calculation of the four basic quantities: A, J, B, C, in descending order, from which the analysis of variance is performed. It will be seen shortly that the analysis of linear models with interactions amounts to adding one more quantity (say, U) between A and J; that is, $A > U > J$.

The linear models for two-way classification without and with interactions are, respectively,

$$y_{ij\alpha} = m + b_i + t_j \qquad + e_{ij\alpha} \tag{1}$$

$$y_{ij\alpha} = m + b_i + t_j + h_{ij} + e'_{ij\alpha} \tag{2}$$

where h_{ij} is the interaction manifested by the observations in the cell of treatment j and block i. These values are called the interactions between blocks and treatments (or between treatments and blocks).

A prime has been added to the e in model (2) to distinguish it from the e in (1). Ordinarily this is unnecessary, as it is always understood that the es of model (1) are not the same as the e in model (2) for the same set of data. However, as we shall establish a relationship between the es of the two models, it is desirable to use a prime to distinguish them. (The prime will be dropped later in the chapter when we talk about model (2) only.)

In the model with interactions, the least square method requires the minimization of the quantity

$$\sum_{ij\alpha} (e'_{ij\alpha})^2 = \sum_{ij\alpha} (y_{ij\alpha} - m - b_i - t_j - h_{ij})^2 \tag{3}$$

Setting the partial derivative of the function above, with respect to a

particular h_{ij} (that is, i and j are both fixed), we obtain

$$\sum_{\alpha} (y_{ij\alpha} - m - b_i - t_j - h_{ij}) = \sum_{\alpha} e'_{ij\alpha} = 0 \tag{4}$$

This shows that the e' of model (2) must add up to zero for each cell (ij). Furthermore, $y_{ij\alpha} - m - b_i - t_j$ is the e-value of model (1) without interactions. What normal equation (4) says is

$$\sum_{\alpha} e_{ij\alpha} = n_{ij} h_{ij}; \quad h_{ij} = \sum_{\alpha} e_{ij\alpha} / n_{ij} \tag{5}$$

where n_{ij} is the number of observations in the cell (ij). The values of $e_{ij\alpha}$ have been found in the last chapter (11). They are reproduced in Table 10.1 to calculate h_{ij} and $e'_{ij\alpha}$. Note that the residual values e in the preceding chapter, although adding up to zero for each row and for each column, do not add up to zero for each cell. The average value of the es in each cell is the interaction h_{ij} (Table 10.1). The deviations of es from their own cell mean are the new error e's which add up to zero for each cell. The sums of squares of e, h, e' are, from Table 10.1,

$$\sum e^2 = 9^2 \quad + (-3)^2 \quad + \cdots + 12^2 \qquad = 750$$

$$\sum h^2 = 2(3)^2 + 4(1.5)^2 + \cdots + 12^2 \qquad = 249$$

$$\sum (e')^2 = 6^2 \quad + (-6)^2 \quad + \cdots + \ 8^2 + 0 = 501$$

Table 10.1. *Values of e, h, e' and their sum of squares*

Values of e from Chapter 9 (11)			Values of h (cell mean of e)			Values of e' (deviations from cell mean)		
9 −3	−2 6 −5 7	−7 1 −6	3 3	1.5 1.5 1.5 1.5	−4 −4 −4	+6 −6	−3.5 +4.5 −6.5 +5.5	−3 +5 −2
5 −12 1	−11 5	12	−2 −2 −2	−3 −3	12	+7 −10 +3	−8 +8	0
$\sum e^2 = 750$ $df = 11$			$\sum h^2 = 249$ $df = 2$			$\sum (e')^2 = 501$ $df = 9$		

In other words, the e of model (1) includes interactions and pure error.

$$e = h + e' \tag{6}$$

$$\sum e^2 = \sum h^2 + \sum (e')^2 \tag{7}$$

$$750 = 249 + 501$$

$$df\colon \quad 11 = 2 + 9$$

where the summation $\sum$ covers all N numbers (i, j, α). In view of (6) and (7), we see model (2) with interactions is merely a subdivision of the e-term of model (1) into two components: interaction h and pure error e'.

The interaction sum of squares has only two degrees of freedom, as h_{ij} adds up to zero for each row and for each column (see middle table of Table 10.1). The pure error (within cell) *ssq* has $\sum_{ij} (n_{ij} - 1) = 15 - 6 = 9$ degrees of freedom.

Practical computation

The procedure of finding the interactions and their sum of squares, presented above, is to illustrate the meaning of interactions. In practice, the e-values of model (1) are never explicitly calculated, because their sum of squares may be obtained by subtraction $(A - J)$. Consequently, the sums of squares for interactions and pure error are never calculated in the fashion of Table 10.1.

Let us examine the errors e' in Table 10.1 once more. In the first cell the deviations are $(+6, -6)$ with mean zero. The *ssq* of these two numbers is the same as that of the two numbers $(9, -3)$ with mean 3. In turn, $(9, -3)$ are originally obtained as $(27 - 18, 15 - 18)$. It follows that the *ssq* of $(+6, -6)$ is the *ssq* of deviations of (27, 15) from their mean 21. To summarize, the values of e' are the deviations of the observed y from their own cell means. This is why we have been calling the e' the 'pure' error, as they are deviations within a cell receiving the same treatment occurring in the same block.

The observations above lead us to calculate the sum of squares within cells as the pure error *ssq*. For this purpose we need the cell totals from the original data in Table 9.1:

Y_{ij} = cell total, (n_{ij}) = cell size (number of observations)

42 (2)	50 (4)	51 (3)
78 (3)	36 (2)	43 (1)

(8)

The six cells are of one-way classification. Calculate the following quantity based on cell totals:

$$U=\sum_{ij}\frac{Y_{ij}^2}{n_{ij}}=\frac{42^2}{2}+\frac{50^2}{4}+\frac{51^2}{3}+\frac{78^2}{3}+\frac{36^2}{2}+\frac{43^2}{1}=6899 \qquad (9)$$

The total sum of the squares may be subdivided as follows ($A=7400$; $C=6000$):

	df	*sum of squares*
Between cells	5	$U-C=$ 899
Within cells	9	$A-U=$ 501
Total	14	$A-C=1400$

(10)

The within-cells *ssq* is the error *ssq*; the between-cells *ssq* includes block effects, treatment effects, and interactions. Since the block and treatment *ssq* have already been obtained in the last chapter, the interaction *ssq* may be obtained by subtraction ($U-J$). The complete analysis of variance is given in Table 10.2.

Table 10.2. *Analysis of variance of data in Table 9.1 with interactions*

	df	*ssq*	*msq*	*F*
Between blocks, ignoring treatments	2	$B-C=$ 321.67	(no test)	
Between treatments, adjusted for blocks	1	$J-B=$ 328.33	328.33	5.90
Interactions	2	$U-J=$ 249.00	124.50	2.24
Error	9	$A-U=$ 501.00	55.67	
Total	14	$A-C=1400.00$		

Summary of calculations

The calculations of the last two chapters may be summarized by introducing the 'indicators' to the basic quantities, as we have done previously in Chapters 3 and 6.

Basis	*Quantities calculated*	
Single observations, $y_{ij\alpha}$	$A = A(m, b, t, h, e) = 7400$	
Cell totals, Y_{ij}	$U = U(m, b, t, h) \quad = 6899$	
Solutions, b_i and t_j	$J \ = J(m, b, t) \quad = 6650$	(11)
Block totals, Y_i	$B = B(m, b) \quad = 6321.67$	
Grand total, Y	$C = C(m) \quad = 6000$	

The successive differences between two adjacent quantities are the *ssq* used in the analysis of variance in Table 10.2. Of the five basic quantities shown above, four (A, U, B, C) may be calculated in a routine manner. Only the quantity J requires the values of b_i and t_j, solutions of the normal equations.

Various other meaningful quantities may be obtained from (11); for example,

$$U - B = ssq(t, h) = 577.33 \tag{12}$$

is the sum of squares due to treatments and interactions (328.33 + 249.00). The arrangement (11) is an abbreviated extension of the scheme exhibited in Table 6.2 without showing the individual numbers.

In order to avoid any possible confusion or misunderstanding about the interactions and orthogonal contrasts (Chapter 8), the following scheme is offered:

- Total
  - Between blocks
  - Within blocks
    - Treatments
      - Subdivision by orthogonal contrasts
    - Error (no interactions)
      - Interactions
      - Pure error

(13)

From this schematic summary it is clear that orthogonal contrasts subdivide the treatment sum of squares, while interaction is a part of the error *ssq* in the model without interactions. Hence, the method of orthogonal contrasts described in Chapter 8 is applicable whether there are interactions or not.

Exercises

1. In Chapter 9 we analyzed the two-way classification data without interaction. In the present chapter we analyzed the same set of

data with interactions. Now we may profitably review the contents of these two chapters by working on the new set of data ($N = 14$) with only two blocks shown below:

	Block I	Block II	Total
Treatment (1)	43, 48, 41	28, 30, 29, 26, 25	$Z_1 = 270$
Treatment (2)	26, 36, 27, 38	7, 16	$Z_2 = 150$

We shall use the linear model with interactions and hence need the cell totals:

	Cell totals Y_{ij};		Cell size	(n_{ij})	
Treatment (1)	132	(3)	138	(5)	
Treatment (2)	127	(4)	23	(2)	
Block total, Y_i	259	(7)	161	(7)	$Y = 420$
Block mean, $\bar{y}_i$	37		23		$\bar{y} = 30$

The first step is to calculate the four basic quantities:

Single values: $A = 43^2 + 48^2 + \cdots + 7^2 + 16^2 = 14\,110$

Cell totals: $U = \dfrac{132^2}{3} + \dfrac{138^2}{5} + \dfrac{127^2}{4} + \dfrac{23^2}{2} = 13\,913.55$

Block totals: $B = \dfrac{259^2}{7} + \dfrac{161^2}{7} = 13\,286$

Grand total: $C = (420)^2/14 = 12\,600$

The second step is to eliminate the block effects b_i from the normal equations by subtracting the corresponding n_{ij} block means from the treatment totals:

$$q_1 = Z_1 - n_{11}\bar{y}_1 - n_{12}\bar{y}_2 = 270 - 3(37) - 5(23) = 44$$
$$q_2 = Z_2 - n_{21}\bar{y}_1 - n_{22}\bar{y}_2 = 150 - 4(37) - 2(23) = -44$$

Since $\sum q_i = 0$, in general, we need to calculate only q_1 when there are two treatments. To obtain the corresponding algebraic expressions for the qs, we replace Z_1 by $8t_1$, replace $\bar{y}_1$ by $(3t_1 + 4t_2)/7$, and replace

$\bar{y}_2$ by $(5t_1+2t_2)/7$. Thus,

$$q_1 = 8t_1 - 3\left(\frac{3t_1+4t_2}{7}\right) - 5\left(\frac{5t_1+2t_2}{7}\right) = 44$$

Simplifying,

$$q_1 = \tfrac{22}{7}(t_1 - t_2) = 44, \quad q_2 = \tfrac{-22}{7}(t_1 - t_2) = -44$$

The third step is to solve for the ts with the help of the constraints

$$8t_1 + 6t_2 = 0, \quad 7b_1 + 7b_2 = 0$$

These constraints lead to $Y = 420 = 14m$, so that $m = \bar{y} = 30$. The solutions are:

$$t_1 = 6, \quad t_2 = -8, \quad \text{and} \quad t_1 - t_2 = 14$$

The fourth step is to calculate the treatment *ssq* (adjusted for blocks):

$$\text{treatment } ssq = \sum q_j t_j = \tfrac{22}{7}(t_1 - t_2)^2 = 616$$

Now we are ready to prepare the table for the analysis of variance. In doing so, we note that Table 10.2 may be slightly shortened by omitting the sum of squares involving the quantity $C = Y^2/N$. The shortened table is as follows:

	df	*ssq*		*df*	*ssq*	*F*
Treatments (adjusted) and interactions	2	$U-B = 627.55$	Treatments	1	616.00	31.36
			Interactions	1	11.55	
Within cells, error	10	$A-U = 196.45$			$s^2 = 19.645$	
Within blocks	12	$A-B = 824.00$				

Since both treatments and interactions have one *df* each, their *ssq* is also the mean square. The treatment differences are significant. The block *ssq* (ignoring treatments) is not shown in the table above. The analysis is based on the sum of squares within blocks. If this procedure of analysis seems long, see the results of Table 11.1 obtained by a shortcut method to be explained in the next chapter.

2. It is instructive and satisfying to find the individual values of interaction (h) for each cell and the error (e) for each observation. Substituting $t_1 = 6$, $t_2 = -8$, and $m = 30$ in the normal equations for block totals,

$$Y_1 = 7(30) + 3(6) + 4(-8) + 7b_1 = 259, \quad 7b_1 = 63, \quad b_1 = 9$$
$$Y_2 = 7(30) + 5(6) + 2(-8) + 7b_2 = 161, \quad 7b_2 = -63, \quad b_2 = -9$$

Then the following table of the values of $m + t + b$ may be obtained:

	$b_1 = 9$	$b_2 = -9$	
$m + t_1 = 36$	45, 45, 45	27, 27, 27, 27, 27	270
$m + t_2 = 22$	31, 31, 31, 31	13, 13	150
	259	161	420

The deviations of the observed ys from these values are $h + e$:

$h + e$

−2 3 −4	1 3 2 −1 −2
−5 5 −4 7	−6 3

$\sum (h+e)^2 = 208$

h

−1 −1 −1	0.6 0.6 0.6 0.6 0.6
0.75 0.75 0.75 0.75	−1.5 −1.5

$\sum h^2 = 11.55$

e

−1 4 −3	0.4 2.4 1.4 −1.6 −2.6
−5.75 4.25 −4.75 6.25	−4.5 4.5

$\sum e^2 = 196.45$

$$\sum (h+e)^2 = \sum h^2 + \sum e^2$$
$$208 = 11.55 + 196.45$$

where the summation covers all fourteen numbers in the table. Note that h is the mean of $h + e$ for each cell, and e is the deviation of $(h+e) - h$, which is also $y_{ij\alpha} - \bar{y}_{ij\cdot}$. These results give a complete check of our calculations of the sum of squares as employed for the analysis of variance.

11

Shortcut method for two treatments

In biomedical research it frequently happens that an experiment involves two treatments only: say, one standard and one new. Some special procedures for such a case have been given in various statistical texts. The special procedures are usually shorter than the general procedure described in the last chapter; but the chief advantage lies in the routine nature of the calculations, which every research worker can follow, there being no normal equations to solve. We shall give one special procedure for comparing two treatments in this chapter, using the same data as before (Chapters 9 and 10). In order to make this chapter self-contained, the original data (Table 9.1) and some preliminary calculations are produced in Table 11.1.

The preliminary calculations shown in Table 11.1 are all of the conventional type (as used in one-way classification) with which the reader is already familiar. In the within-cells line, $ssq/df = 501/9 = 55.67$ provides an estimate of the error variance (for linear models with interactions).

The between-cells *ssq* is $U - C = 899$ with 5 degrees of freedom. It contains three components: namely, block effects, treatment effects, and interactions. One component is the *ssq* due to blocks, ignoring treatments, $B - C = 321.67$, so that the sum of the other two components is $U - B = 577.33$. The latter contains the treatment *ssq* (adjusted for blocks) with $df = 1$ and interaction *ssq* with $df = 2$. Our subsequent task is to find the treatment *ssq* and then the interaction *ssq* may be obtained by subtraction.

Treatment sum of squares

The special method of finding the treatment *ssq* (adjusted for blocks) for two treatments is really brief and straightforward, and that

Table 11.1. *Two-way classification data (from table 9.1) and preliminary calculations (note that the (n_{ij}) are in parentheses to distinguish them from the y values,*

	Block I	Block II	Block III	
Treatment (1)	27 15	9 17 6 18	14 22 15	
Treatment (2)	33 16 29	10 26	43	
Cells totals, Y_{ij}	42 (2)	50 (4)	51 (3)	
	78 (3)	36 (2)	43 (1)	
Block totals, Y_j	120 (5)	86 (6)	94 (4)	300 (15)

Single values: $A = 27^2 + 15^2 + 9^2 + \cdots + 26^2 + 43^2 \quad = 7400.00$

Cell totals: $U = \frac{42^2}{2} + \frac{50^2}{4} + \frac{51^2}{3} + \frac{78^2}{3} + \frac{36^2}{2} + \frac{43^2}{1} \quad = 6899.00$

Block totals: $B = \frac{120^2}{5} + \frac{86^2}{6} + \frac{94^2}{4} \quad = 6321.67$

Grand total: $C = (300)^2/15 \quad = 6000.00$

	df	*ssq*		*df*	*ssq*
Between cells					
Between blocks, ignoring treatments	2	$B - C = 321.67$			
Treatments (adjusted) and interactions	3	$U - B = 577.33$	Treatments	1	see (4) (5) (6)
			Interactions	2	
Within cells (error)	9	$A - U = 501.00$	$s^2 = 55.67$		
Total	14	$A - C = 1400.00$			

is why it is worth knowing. We shall give the method first, leaving its justification to the last section of the chapter.

The method is based on cell *means* calculated from the cell totals in Table 11.1. The cell means are given in the upper portion of Table 11.2, with the (n_{ij}) indicated in parentheses. The first step is to take the differences between the cell means of the same block. In the table we have used

$$d_j = \bar{y}_{1j} - \bar{y}_{2j} \qquad (1)$$

so that, for the first block, $d_1 = 21 - 26 = -5$. But, the designation of treatments (1) and (2) is arbitrary. If we define $d_1 = \bar{y}_{21} - \bar{y}_{11} = 26 - 21 = 5$, it does not matter, as the sum of squares depends on the square of these differences. Whichever way we choose to define d_j, it must be followed consistently in all the blocks.

The second step is to assign a 'weight' (or coefficient) w_j to each d_j, where

$$w_j = \frac{n_{1j} n_{2j}}{n_{1j} + n_{2j}} \qquad (2)$$

Table 11.2. *Calculation of treatment sum of squares (adjusted for blocks) and interaction sum of squares, using cell means*

	I	II	III	
Mean, $\bar{y}_{1j}$	21 (2)	12.5 (4)	17 (3)	
Mean, $\bar{y}_{2j}$	26 (3)	18 (2)	43 (1)	
$d_j = \bar{y}_{1j} - \bar{y}_{2j}$	−5.0	−5.5	−26.0	$\bar{d} = -10$
Weight, w_j:	$\frac{6}{5}$	$\frac{8}{6}$	$\frac{3}{4}$	$\sum w_j = \frac{197}{60} = 3.283$
$w_j d_j$:	−6.00	−7.33	−19.50	$\sum w_j d_j = -32.833$
$w_j d_j^2$:	30.00	40.33	507.00	$\sum w_j d_j^2 = 577.333$

ssq (treatments and interactions) $= \sum w_j d_j^2 = 577.333$

ssq (treatments, adjusted for blocks) $= (\sum w_j d_j)^2 / \sum w_j = 328.333$

Subtracting, *ssq* (interactions) $= \sum w_j d_j^2 - (\sum w_j d_j)^2 / \sum w_j = 249.000$

These weights are easily calculated from the (n_{ij}) associated with each cell mean.

The third step is to calculate the products $w_j d_j$ and their sum (Table 11.2). The weighted average of the d_js is

$$\bar{d} = \frac{\sum w_j d_j}{\sum w_j} = \frac{-32.833}{3.283} = -10 \tag{3}$$

The treatment *ssq* (adjusted for blocks) is then

$$\text{treatment } ssq = (\textstyle\sum w_j)\bar{d}^2 \quad = (3.283)(-10)^2 = 328.33 \tag{4}$$

or

$$= \frac{(\sum w_j d_j)^2}{\sum w_j} \quad = \frac{(-32.833)^2}{3.283} \quad = 328.33 \tag{4'}$$

The interaction *ssq* is then $577.33 - 328.33 = 249.00$. Putting these *ssq* in the lower right of Table 11.1 completes the analysis of variance table. The result is, of course, the same as Table 10.2, to which the reader should refer for the *F*-tests.

Interaction sum of squares

For most practical purposes, the calculations in the preceding section are all we need to complete the analysis of variance. If desired, however, the interaction *ssq* may be calculated independently instead of by subtraction. Let us examine Table 11.2 once more. If there is no interaction between blocks and treatments, the differences d_j between the two treatment *means* should be the same for all blocks. But the observed d_j varies from block to block, indicating the presence of interaction. The interaction *ssq* is given by the sum of squares of these *d*s, namely,

$$\begin{aligned} &\text{interaction } ssq \\ &\quad = \textstyle\sum w_j (d_j - \bar{d})^2 \\ &\quad = \tfrac{6}{5}(-5+10)^2 + \tfrac{8}{6}(-5.5+10)^2 + \tfrac{3}{4}(-26+10)^2 = 249 \end{aligned} \tag{5}$$

correctly. The expression (5) may also be written

$$\text{interaction } ssq = \textstyle\sum w_j d_j^2 - \frac{(\sum w_j d_j)^2}{\sum w_j} \tag{5'}$$

Note that the second term is the treatment *ssq* (4′). So, the first term must be the sum of squares due to treatments and interactions. Table 11.2 shows that this is indeed the case, noting

$$\textstyle\sum w_j d_j^2 = 577.33 = U - B \tag{4)+(5}$$

which is $U-B$ in Table 11.1. The interaction *ssq* is then $577.33-328.33=249$, in agreement with (5).

When the interaction *ssq* is calculated independently, as was done in Table 11.2, the full analysis of variance table (Table 10.2) may be shortened somewhat. There is no longer any need to calculate the quantities B and C, based on block and grand totals, although we still need A and U to obtain the within-cell sum of squares. The shortened table consists of only three relevant lines:

		df	*sum of squares*	*msq*	*F*	
$U-B$	Treatments	1	$(\sum w_j)\bar{d}^2=328.33$	328.33	5.90	(6)
	Interactions	2	$\sum w_j(d_j-\bar{d})^2=249.00$	124.50	2.24	
$A-U$	Error	9	$A-U=501.00$	55.67		

The tabulated value of F is 5.12 at $P=0.05$ level for (1,9) degrees of freedom, so the treatment difference is barely significant at the 5% level. The interactions are non-significant.

Correspondence with linear model

The shortcut exhibited in Table 11.2 is entirely consistent with the linear model $y=m+t+b+h+e$ introduced in Chapter 10(2). The algebraic demonstration of the correspondence is simple and straightforward. Since the es add up to zero within each cell, the cell means do not involve the es. In the jth block, the two cell means are

$$\bar{y}_{1j}=m+t_1+b_j+h_{1j}$$
$$\bar{y}_{2j}=m+t_2+b_j+h_{2j}$$

so that

$$d_j=\bar{y}_{1j}-\bar{y}_{2j}=t_1-t_2+h_{1j}-h_{2j} \qquad (7)$$

and

$$\sum w_jd_j=(\sum w_j)(t_1-t_2)+\sum w_j(h_{1j}-h_{2j}) \qquad (8)$$

The key relationship to be demonstrated is $\sum w_j(h_{1j}-h_{2j})=0$. Recall that the interaction values h_{ij} add up to zero for each treatment (row) and for each block (column), as shown in Table 10.1, middle. Thus, for the first row,

$$\sum_j n_{1j}h_{1j}=n_{11}h_{11}+n_{12}h_{12}+n_{13}h_{13}=0 \qquad (9)$$

and for each block (fixed j),

$$n_{1j}h_{1j}+n_{2j}h_{2j}=0, \quad h_{2j}=\frac{-n_{1j}}{n_{2j}}h_{1j} \tag{10}$$

so that

$$h_{1j}-h_{2j}=h_{1j}+\frac{n_{1j}h_{1j}}{n_{2j}}=\frac{n_{1j}+n_{2j}}{n_{2j}}h_{1j} \tag{11}$$

Hence,

$$\sum_j w_j(h_{1j}-h_{2j})=\sum_j \left(\frac{n_{1j}n_{2j}}{n_{1j}+n_{2j}}\right)\left(\frac{n_{1j}+n_{2j}}{n_{2j}}\right)h_{1j}$$

$$=\sum_j n_{1j}h_{1j}=0 \tag{12}$$

by (9). And from (8), we obtain the key relationship:

$$\sum w_j d_j=(\sum w_j)(t_1-t_2) \tag{13}$$

$$\bar{d}=\frac{\sum w_j d_j}{\sum w_j}=t_1-t_2 \tag{14}$$

In our example, $d=-10$, which agrees with the solution $t_1=-4$, $t_2=6$ of the normal equations in Chapter 9(5s). Also, from Chapter 9(6′),

$$\text{treatment } ssq=\sum_i q_i t_i=\left(\sum_j w_j\right)(t_1-t_2)^2$$

$$=\left(\sum_j w_j\right)\bar{d}^2, \quad \text{that is,} \quad (4)$$

Since the deviation $d_j-\bar{d}=h_{1j}-h_{2j}$ is a function of h only, further algebra (exercise) making use of (10) and (11) will establish

$$\sum_j w_j(d_j-\bar{d})^2=\sum_j w_j(h_{1j}-h_{2j})^2=\sum_{ij} n_{ij}h_{ij}^2 \tag{15}$$

which is the interaction sum of squares. It is also clear that $\sum w_j d_j^2$ contains both treatment and interaction sum of squares, as the sum of products $\sum w_j(t_1-t_2)(h_{1j}-h_{2j})=0$.

Shortcut for $2\times k$ classification

Although the example in this chapter consists of two treatments, it is immediately seen that the shortcut method may also be used when there are $k>2$ treatments but only two blocks. In Chapter 5 it was

pointed out that, for arithmetic convenience in the process of solving the normal equations, we always eliminate the effects of the factor with the larger number of classes and solve for the effects of the factor with the smaller number of classes. Here is another application of that principle.

Suppose that the three columns of Table 11.1 represent three treatments of interest and that the two rows are the two blocks (the dual case). In such a case we still do the same calculation as indicated in Table 11.2; and then apply the general relationship

$$\begin{aligned}\text{joint } ssq&(\text{rows and columns})\\ &= ssq(\text{columns, ignoring rows}) + ssq(\text{rows, adjusted})\\ &= ssq(\text{columns, adjusted}) + ssq(\text{rows, ignoring columns})\end{aligned}$$

while the error *ssq* and the interaction *ssq* will remain the same in both cases. In our numerical example it has already been found that

$$\begin{aligned} ssq(\text{columns, ignoring rows}) &= 321.67 \text{ (Table 11.1)}\\ ssq(\text{rows, adjusted}) &= 328.33 \text{ (Table 11.2)}\\ \text{joint } ssq(\text{rows and columns}) &= 650.00\end{aligned}$$

To obtain the *ssq*(columns, adjusted), we need only to calculate the *ssq* (rows, ignoring columns). The latter is, from the row totals in Table 11.1,

$$\frac{143^2}{9}+\frac{157^2}{6}-\frac{300^2}{15}=380.28$$

so that the desired

$$ssq(\text{columns, adjusted}) \text{ is } 650-380.28=269.72$$

with 2 *df*, and the corresponding mean square is $269.72/2 = 134.86$. The appropriate test is given by $F = 134.86/55.67 = 2.42$ with (2, 9) degrees of freedom, being non-significant.

In summary we may say that the shortcut method described in this chapter is applicable for all unbalanced data with $2\times k$ classifications whether there are two treatments or two blocks.

Exercises

1. The shortcut procedure for two treatments applies to any $2\times k$ table, where k is the number of blocks. The reader may practice the method on the following table with four blocks (I, II, III, IV) and two treatments (1) and (2).

Individual observed values

	I	II	III	IV
(1)	27 15	9 18 6	17	14 22 15
(2)	33 29	16	10 26	43
Y_j:	104	49	53	94

Cell total, cell mean, (n_{ij})

	I	II	III	IV
(1)	42 21 (2)	33 11 (3)	17 17 (1)	51 17 (3)
(2)	62 31 (2)	16 16 (1)	36 18 (2)	43 43 (1)
Y_j:	104 (4)	49 (4)	53 (3)	94 (4)

Hint: Always do the preliminary calculations:

$$A = \sum_{ij\alpha} y_{ij\alpha}^2 = 7400$$

$$U = \sum_{ij} \frac{Y_{ij}^2}{n_{ij}} = 7076$$

$$B = \sum_{j} \frac{Y_j^2}{n_j} = 6449.583$$

$$C = Y^2/N = 6000$$

$A - U = 324.000$ within cells

$U - B = 626.417$ treatments and interactions

$B - C = 449.583$ between blocks, ignoring treatments

Also note $A - B = 950.417 = ssq$ within blocks, ignoring treatments, and $U - C = 1076 = ssq$ between the eight cells. The estimate of the error variance is $s^2 = (A - U)/\sum (n_{ij} - 1) = 324/7 = 46.2857$.

Answer: Follow the procedure outlined in Table 11.2. The purpose is to find the sum of squares due to treatments and that due to interactions, that is, a subdivision of $U - B$. Do as much as you can and then check your results with those given in Chapter 12. Look for them.

2. Prove (15):

$$\text{interaction } ssq = \sum w_j (d_j - \bar{d})^2 = \sum w_j (h_{1j} - h_{2j})^2$$

where j indicates $(j = 1, 2, \ldots, k)$.

Proof: Substituting (2), (10), and (11) in (15), the latter becomes

$$\text{interaction } ssq = \sum_j \frac{n_{1j}n_{2j}}{n_{1j}+n_{2j}}\left(\frac{n_{1j}+n_{2j}}{n_{2j}}\right)^2 h_{1j}^2$$

$$= \sum_j n_{1j}\left(1+\frac{n_{1j}}{n_{2j}}\right)h_{1j}^2$$

$$= \sum_j n_{1j}h_{1j}^2 + \sum_j n_{2j}\left(\frac{n_{1j}}{n_{2j}}\right)^2 h_{1j}^2$$

$$= \sum_j n_{1j}h_{1j}^2 + \sum_j n_{2j}h_{2j}^2 = \sum_{ij} n_{ij}h_{ij}^2$$

12

Factors at two levels

In biomedical research it frequently happens that the 'blocks' represent another factor of interest. When two factors (say, a and b) of interest are investigated in an experiment, then the data are said to be factorial. The simplest example of data of this nature is the case in which there are only two levels for each factor: that is, a_1, a_2 for factor a and b_1, b_2 for factor b, so that there are four (2×2) treatment groups (Table 12.1). This chapter deals with the analysis of such simple factorial data.

The shortcut analysis

The general procedure is to analyze and test one factor at a time. Suppose we want to test the effects of factor a first. Then factor b is taken as 'blocks' (for which no test is to be made). The dual analysis (Chapter 5), using factor a as blocks, will test the effects of factor b.

Let us first regard a_1 and a_2 as two treatments, b_1 and b_2 playing the role of two blocks. Since it has already been shown that the shortcut method for comparing two treatments (Chapter 11) is equivalent to the analysis of linear models with interactions, we may adopt that shortcut method here in comparing a_1 and a_2 without further ado. In Table 12.1 the following quantities for each block j and their sums (over the blocks) are calculated:

$$d_j = \bar{y}_{1j} - \bar{y}_{2j}, \quad w_j = \frac{n_{1j} n_{2j}}{n_{1j} + n_{2j}}, \quad w_j d_j, \quad w_j d_j^2 \tag{1}$$

The average difference between cell means is (Table 12.1)

$$\bar{d} = \frac{\sum w_j d_j}{\sum w_j} = \frac{44 \times 7}{22} = 14.00 \tag{2}$$

The various sums of squares (ssq) relevant to the analysis of variance are calculated as follows:

$df = 2$, treatments and interactions $\quad \sum w_j d_j^2 = 627.55$

$df = 1$, treatments (adjusted for blocks) $\quad (\sum w_j)\bar{d}^2 = 616.00 \quad (3)$

$df = 1$, interactions, $\sum w_j(d_j - \bar{d})^2 = \sum w_j d_j^2 - (\sum w_j)\bar{d}^2 = 11.55$

For linear models with interactions, the sum of squares within cells is the error ssq for the analysis of variance. With Y_{ij} = cell total, we calculate from Table 12.1:

$$df = 14, \quad A = \sum_{ij\alpha} y_{ij\alpha}^2 = 43^2 + 48^2 + \cdots + 7^2 + 16^2 = 14\,110.00$$

$$df = 4, \quad U = \sum_{ij} \frac{Y_{ij}^2}{n_{ij}} = \frac{132^2}{3} + \frac{138^2}{5} + \frac{127^2}{4} + \frac{23^2}{2} = 13\,913.55$$

Table 12.1. *Analysis of* 2×2 *factorial data*

		Factor b as blocks	
		b_1	b_2
Factor a as treatments	a_1	43, 48, 41	28, 30, 29, 26, 25
	a_2	26, 36, 27, 38	7, 16

(n_{ij}) = number of observations in a cell

Cell total, Y_{1j} Cell mean, $\bar{y}_{1j}$	132 44.00 (3)	138 27.60 (5)	
Cell total, Y_{2j} Cell mean, $\bar{y}_{2j}$	127 31.75 (4)	23 11.50 (2)	
$\bar{y}_{1j} - \bar{y}_{2j} = d_j$:	12.25	16.10	$\bar{d} = 14.00$
Weight, w_j:	12/7	10/7	$\sum w_j = 22/7$
$w_j d_j$:	21.00	23.00	$\sum w_j d_j = 44.00$
$w_j d_j^2$:	257.25	370.30	$\sum w_j d_j^2 = 627.55$
			Treatment $ssq = (\sum w_j)\bar{d}^2 = 616.00$
			Subtracting (interactions) = 11.55

The error variance is estimated by

$$s^2 = \frac{A-U}{14-4} = \frac{196.45}{10} = 19.645 \qquad (4)$$

The variance-ratio for treatments is $F = 616/19.645 = 31.36$ with (1, 10) degrees of freedom. The treatment effects are significant. These results are identical with those obtained in the exercises at the end of Chapter 10, wherein the long formal method of finding t_1, t_2, and $\sum q_i t_i$ was used.

The results of the dual analysis is given in Table 12.2, in which the two treatments are b_1 and b_2 while the two blocks are a_1 and a_2. In practice, the original measurements need not be reproduced; all we need is a rearrangement of the cell means. Nevertheless, they are shown to make the table self-contained.

Comparing Tables 12.1 and 12.2, we see that their d_j and w_j are different, and so are $w_j d_j$ and $w_j d_j^2$. But the interaction $ssq = 11.55$ remains the same, as expected from our studies on the joint sum of

Table 12.2. *Dual analysis of data in Table* 12.1

		Factor a as blocks		
		a_1	a_2	
Factor b as treatments	b_1	43, 48, 41	26, 36, 27, 38	
	b_2	28, 30, 29, 26, 25	7, 16	
Cell mean,	$\bar{y}_{1j}$	44.00 (3)	31.75 (4)	
	$\bar{y}_{2j}$	27.60 (5)	11.50 (2)	
	d_j:	16.40	20.25	$\bar{d} = 18.00$
	w_j:	15/8	8/6	$\sum w_j = 77/24$
	$w_j d_j$:	30.75	27.00	$\sum w_j d_j = 57.75$
	$w_j d_j^2$:	504.30	546.75	$\sum w_j d_j^2 = 1051.05$
				Treatment $ssq = (\sum w_j)\bar{d}^2 = 1039.50$
				Subtracting (interactions) $= 11.55$

squares for blocks and treatments (Chapter 6). The error variance $s^2 = 19.645$ also remains the same, of course. The variance-ratio for testing the factor b effects is $F = 1039.50/19.645 = 52.91$ with (1, 10) degrees of freedom.

As a final check on the relationships of the two sets of results, we may calculate the sum of squares between blocks, ignoring treatments, from Tables 12.1 and 12.2, noting $C = Y^2/N = (420)^2/14 = 12\,600$ for both tables.

In Table 12.1			*In Table* 12.2		
$B_1 = \frac{259^2}{7} + \frac{161^2}{7}$	=	13 286.00	$B_2 = \frac{270^2}{8} + \frac{150^2}{6}$	=	12 862.50
Blocks, $B_1 - C$	=	686.00	Blocks, $B_2 - C$	=	262.50
Treatments, adjusted	=	616.00	Treatments, adjusted	=	1039.50
Joint *ssq*	=	1302.00	Joint *ssq*	=	1302.00

(5)

Method of orthogonal contrasts

We shall now introduce another method of analyzing the factorial data which may be readily extended to the case of three factors ($2 \times 2 \times 2$). The method of orthogonal contrasts has been explained in many texts, but mostly as applied to orthogonal data (equal numbers of replications for all treatments). Here we shall apply the same principle of orthogonal contrasts to cases with an unequal number of replications, using the data in Table 12.2 as an illustration.

It is convenient to arrange the four treatment cells in a single file (instead of the 2×2 formation in Table 12.2). When arranged in a single file, it is also convenient to denote the number of observations in a cell simply by n_i rather than by n_{ij}. Similarly, the cell totals may be denoted by Y_i instead of Y_{ij}. If we use factor a as blocks and factor b as treatments (Table 12.2), then the single-file arrangement of the four cells would be (a_1b_1), (a_1b_2), (a_2b_1), (a_2b_2), as indicated in Table 12.3. Also, the block size is indicated by $N_{12} = n_1 + n_2$ for the a_1-block and $N_{34} = n_3 + n_4$ for the a_2-block.

In the lower half of Table 12.3, each row gives the coefficients for a particular contrast. The notation, although conventional in the analysis of factorial data, needs a few words of explanation to avoid any possible confusion with previous symbols. The first contrast, denoted by (A), is

a comparison between a_1 and a_2 as blocks, ignoring the factor b as treatments. This letter (A), meaning comparison (A), is not to be confused with $A = \sum y_{ij\alpha}^2$ employed previously in this book.

The second contrast, denoted by (B) in Table 12.3, gives comparisons between b_1 and b_2 *within* each block; that is, adjusted for blocks. Again, this letter (B), meaning comparison (B), is not to be confused with $B = \sum Y_j^2/n_j$ used previously.

The sequence of classification is important in the analysis of factorial data. In Table 12.3 the first classification is according to factor a and the second classification is according to factor b *within* the classification of a. So we shall adopt the convention that the first contrast, such as (A), implies that it ignores the factor b that appears below it. Conversely, the second contrast, such as (B), implies that it has been adjusted for the factor a that appears above it. We shall always write the contrasts in that order.

Table 12.3. *Subdivision of sum of squares between cells by orthogonal contrasts for* 2×2 *factorial data*

Factor a (as blocks)		a_1		a_2	
Factor b (as treatments)		b_1	b_2	b_1	b_2
Observed values (y)		⋮	⋮ ⋮	⋮ ⋮	⋮
Cell total		Y_1	Y_2	Y_3	Y_4
Cell size		n_1	n_2	n_3	n_4
Block size		$N_{12} = n_1 + n_2$		$N_{34} = n_3 + n_4$	
Between as, ignoring bs	(A)	$\frac{1}{N_{12}}$	$\frac{1}{N_{12}}$	$\frac{-1}{N_{34}}$	$\frac{-1}{N_{34}}$
Between bs, adjusted for as	(B)	$\frac{n_2}{N_{12}}$	$\frac{-n_1}{N_{12}}$	$\frac{n_4}{N_{34}}$	$\frac{-n_3}{N_{34}}$
Interactions	(AB)	$\frac{1}{n_1}$	$\frac{-1}{n_2}$	$\frac{-1}{n_3}$	$\frac{1}{n_4}$

The third and last contrast, denoted by (AB) in Table 12.3, gives the interaction between factors a and b. The interaction, as we learned in previous chapters (particularly in Tables 12.1 and 12.2), remains the same whether the sequence of classification is a first and then b, or b first and then a.

Now, a few words about the general features of the coefficients for the various contrasts. Let k_1, k_2, k_3, k_4 be the coefficients in any row in the lower half of Table 12.3. Then $\sum n_i k_i = 0$. In fact, this is the definition of a contrast or comparison. Furthermore, if k_i' are the coefficients of another row, then $\sum n_i k_i k_i' = 0$. This is the definition of orthogonality of the various contrasts. For practical application, the coefficients of any row may be multiplied by a constant without destroying the properties described above. In particular, the coefficients in any row may change sign (that is, multiplied by -1). So, the coefficients shown in Table 12.3 may be written in many different ways, depending on arithmetic convenience in practical application.

Once the coefficients for the various contrasts are known, the calculation of the various sum of squares is then straightforward. For each row of coefficients k_i, we calculate the following quantities:

$$L = \sum k_i Y_i \tag{6}$$

$$D = \sum k_i^2 n_i \tag{7}$$

and

$$ssq = L^2/D \tag{8}$$

Each such sum of squares has one degree of freedom. Table 12.4 gives a numerical example, using the same data of Table 12.2. For example, the contrast (B) yields

$$L = \tfrac{5}{8}(132) - \tfrac{3}{8}(138) + \tfrac{2}{6}(127) - \tfrac{4}{6}(23) = 57.75 \tag{6'}$$

with denominator

$$D = 3(\tfrac{5}{8})^2 + 5(\tfrac{3}{8})^2 + 4(\tfrac{2}{6})^2 + 2(\tfrac{4}{6})^2 = \tfrac{77}{24} \tag{7'}$$

so that the sum of squares due to factor b, adjusted for factor a, is

$$ssq = L^2/D = (57.75)^2 24/77 = 1039.50 \tag{8'}$$

in agreement with that obtained in Table 12.2. Likewise, the interaction (AB) $ssq = 11.55$ also agrees with the result in Table 12.2.

The sum of squares due to contrast (A) is 262.50 which is the same as $B_2 - C$ given in (5). The sum of the three ssq in Table 12.4 is 1313.55 which is $U - C$, the sum of squares between cells.

The remaining calculations and F-test remain the same as before, viz., $s^2 = (A - U)/(14-4) = 196.45/10 = 19.645$. But there is no test for contrast (A) (between blocks, ignoring treatments).

A few lines of algebra will establish the equivalence between the method of orthogonal contrasts and the shortcut method given earlier. Briefly, the first contrast (A) reduces to $L = \bar{y}_{a1} - \bar{y}_{a2}$, where $\bar{y}_{a1}$ is the mean of block a_1, etc. The corresponding sum of squares is

$$\frac{L^2}{D} = \frac{N_{12}N_{34}}{N_{12}+N_{34}}(\bar{y}_{a1} - \bar{y}_{a2})^2$$

which is the *ssq* between the two blocks.

Similarly, it may be shown that the second contrast (B) reduces to $L = w_1(\bar{y}_1 - \bar{y}_2) + w_2(\bar{y}_3 - \bar{y}_4) = \sum w_j d_j = 57.75$ in Tables 12.2 and 12.4, and D reduces to $\sum w_j$, so that the sum of squares L^2/D reduces to $(\sum w_j d_j)^2/\sum w_j = (\sum w_j)\bar{d}^2$. Finally, the interaction (AB) reduces to

$$L = \bar{y}_1 - \bar{y}_2 - \bar{y}_3 + \bar{y}_4 = h_{11} - h_{12} - h_{21} + h_{22}$$

and L^2/D may be shown to be the *ssq* due to interaction by the same method we used in Chapter 11.

Table 12.4. *Subdivision of sum of squares between cells by orthogonal contrasts. Data from Table* 12.2; *coefficients from Table* 12.3

	a_1		a_2				
	b_1	b_2	b_1	b_2			
Y_i	132	138	127	23	$\sum k_i Y_i$	$\sum k_i^2 n_i$	*ssq*
n_i	3	5	4	2	L	D	L^2/D
(A)	$\frac{1}{8}$	$\frac{1}{8}$	$\frac{-1}{6}$	$\frac{-1}{6}$	8.75	$\frac{7}{24}$	262.50
(B)	$\frac{5}{8}$	$\frac{-3}{8}$	$\frac{2}{6}$	$\frac{-4}{6}$	57.75	$\frac{77}{24}$	1039.50
(AB)	$\frac{1}{3}$	$\frac{-1}{5}$	$\frac{-1}{4}$	$\frac{1}{2}$	−3.85	$\frac{77}{60}$	11.55

Sum of squares between cells $= U - C = 1313.55$

Three factors, each at two levels

The method of orthogonal contrasts may be extended to three factors each with two levels. Table 12.5 gives the coefficients for the various comparisons when the classification of the three factors is in the sequence *a*, *b*, *c*. As in the previous case, the eight cells are arranged in a single file so that a simpler notation (for example, n_i, Y_i) may be adopted. Also, as before, $N_{12}=n_1+n_2$, $N_{1234}=n_1+n_2+n_3+n_4$, etc.

In each row corresponding to each contrast, there are eight coefficients $(k_1, k_2, \ldots, k_8)$. When the coefficients have the same denominator, the latter is typed only once in Table 10.5. Thus,

$$\frac{n_2 \qquad -n_1}{N_{12}} \quad \text{means} \quad \frac{n_2}{N_{12}}, \quad \frac{-n_1}{N_{12}}$$

For interactions (AC) and (BC), it is found that the coefficients may be more conveniently expressed in terms of the reciprocals $w_i = 1/n_i$ and

$$W_{12}=w_1+w_2=\frac{1}{n_1}+\frac{1}{n_2}$$

$$W_{1234}=w_1+w_2+w_3+w_4=\frac{1}{n_1}+\frac{1}{n_2}+\frac{1}{n_3}+\frac{1}{n_4}$$

That is, the capital W is a sum of reciprocals, not the reciprocal of a sum.

As remarked previously, the coefficients in each row may be multiplied by a constant without destroying the orthogonality of the contrast with others or affecting the sum of squares L^2/D. Hence, for numerical computation, the investigator is free to adjust these coefficients to convenient numbers.

As a numerical example we may use the data in Exercise 1 of Chapter 11, taking treatments (1) and (2) as c_1 and c_2 and the four blocks as I $=a_1b_1$, II $=a_1b_2$, III $=a_2b_1$, IV $=a_2b_2$. The data will take the following form, when arranged in a single file:

	I		II		III		IV			
	c_1	c_2	c_1	c_2	c_1	c_2	c_1	c_2		
Cell total, Y_i:	42	62	33	16	17	36	51	43	$Y=300$	(9)
Cell size n_i:	2	2	3	1	1	2	3	1	$N=$ 15	

Meager data like this are probably not worth analyzing in a factorial manner but here we are using them to illustrate the procedure without

Table 12.5. *Subdivision of sum of squares between cells by orthogonal contrasts for* 2×2×2 *data. Each contrast ignores the factors below it but is adjusted for the factors above it* (*C. C. Li & S. Mazumdar* (1976), *Journal of Chronic Diseases*, **29**, 355–70)

Factor a	a_1				a_2			
Factor b	b_1		b_2		b_1		b_2	
Factor c	c_1	c_2	c_1	c_2	c_1	c_2	c_1	c_2
Total, Y_i	Y_1	Y_2	Y_3	Y_4	Y_5	Y_6	Y_7	Y_8
Size, n_i	n_1	n_2	n_3	n_4	n_5	n_6	n_7	n_8
(A)	1	1	1	1	−1	−1	−1	−1
	N_{1234}				N_{5678}			
(B)	N_{34}	N_{34}	$-N_{12}$	$-N_{12}$	N_{78}	N_{78}	$-N_{56}$	$-N_{56}$
	N_{1234}				N_{5678}			
(AB)	1	1	−1	−1	−1	−1	1	1
	N_{12}		N_{34}		N_{56}		N_{78}	
(C)	n_2	$-n_1$	n_4	$-n_3$	n_6	$-n_5$	n_8	$-n_7$
	N_{12}		N_{34}		N_{56}		N_{78}	
(AC)	$w_1 W_{34}$	$-w_2 W_{34}$	$w_3 W_{12}$	$-w_4 W_{12}$	$-w_5 W_{78}$	$w_6 W_{78}$	$-w_7 W_{56}$	$w_8 W_{56}$
	W_{1234}				W_{5678}			
(BC)	w_1	$-w_2$	$-w_3$	w_4	w_5	$-w_6$	$-w_7$	w_8
	W_{1234}				W_{5678}			
(ABC)	w_1	$-w_2$	$-w_3$	w_4	$-w_5$	w_6	w_7	$-w_8$

excessive arithmetic. Substituting these n_i or $w_i = 1/n_i$ values in Table 12.5, we obtain the coefficients and then L, D, and *ssq* according to (6), (7), (8). We leave the arithmetic details as an exercise. The resulting sum of squares, each with one degree of freedom, are as follows:

Contrast	*ssq*	*Subtotal*	*df*	*Remark*	
(A)	13.125				
(B)	82.452	449.583	3	Between blocks, ignoring factor c	
(AB)	354.006				
(C)	363.265	363.265	1	Main effect of c, adjusted for blocks	(10)
(AC)	31.849				
(BC)	57.109	263.152	3	Interactions of c with other factors	
(ABC)	174.194				
Total	$U-C = 1076.000$		7	Between cells	

The first three $ssq = msq$ (because $df = 1$) are *not* to be tested against error variance $s^2 = 46.2857$, because they ignore the factor c. Each of the remaining four $ssq = msq$, however, may be tested against error variance s^2, as they have been adjusted for blocks (the other two factors). In other words, the classification of factors according to the sequence of a, b, c, and its subsequent analysis by the method of Table 12.5, yield tests of the effect of factor c and its interactions with the other two factors. The factor to be tested is always the last factor in the sequence of classification.

When we wish to test the effects of factor a or b, the sequence of classification has to be made accordingly. There are six possible sequences of classification for three factors; and they fall into three pairs as shown in Table 12.6. For each pair of sequences, the relationship between the sum of squares for the various contrasts is indicated. Thus, for the first pair of sequences, the sum $(AC)+(BC)$ of the first sequence is equal to the sum $(BC)+(AC)$ of the second sequence. The three-factor interaction (last row of Table 12.6) remains the same for all six sequences of classification.

Pooling the interactions

In (10) we have found the *ssq* for each of the seven contrasts. This is the ultimate analysis for $2\times2\times2$ factorial data. In many practical cases, however, this may not be necesssary, especially in view of the difficulty in interpreting the interactions (AC), (BC), (ABC). If we were willing to pool these three interactions into one category as 'interactions with other factors', the calculations may be much simplified, as we may then use the shortcut method for two treatments instead of the method of orthogonal contrasts. As a numerical example, let us again use the data in Exercise 1 of Chapter 11.

Cell mean	I	II	III	IV		
$\bar{y}_{1j}$	21 (2)	11 (3)	17 (1)	17 (3)		(11)
$\bar{y}_{2j}$	31 (2)	16 (1)	18 (2)	43 (1)		
d_j	10	5	1	26	$\bar{d}=$	10.710 526
w_j	1	3/4	2/3	3/4	$\sum w_j=$	19/6
w_jd_j	10	3.75	0.6667	19.50	$\sum w_jd_j=$	33.916 667
$w_jd_j^2$	100	18.75	0.6667	507.0	$\sum w_jd_j^2=$	626.417

Check with (10), $C\ ssq=(\sum w_j)\bar{d}^2=363.265$

Check with (10), $((AC), (BC), (ABC))$, subtracting $=263.152$

Table 12.6. *The six possible sequences of classification of three factors (a, b, c) and the relationship between the sum of squares of the various contrasts*

a b c		b a c		a c b		c a b		b c a		c b a
$\left.\begin{matrix}(A)\\(B)\end{matrix}\right\}$	=	$\left\{\begin{matrix}(B)\\(A)\end{matrix}\right.$		$\left.\begin{matrix}(A)\\(C)\end{matrix}\right\}$	=	$\left\{\begin{matrix}(C)\\(A)\end{matrix}\right.$		$\left.\begin{matrix}(B)\\(C)\end{matrix}\right\}$	=	$\left\{\begin{matrix}(C)\\(B)\end{matrix}\right.$
(AB)	=	(BA)		(AC)	=	(CA)		(BC)	=	(CB)
(C)	=	(C)		(B)	=	(B)		(A)	=	(A)
$\left.\begin{matrix}(AC)\\(BC)\end{matrix}\right\}$	=	$\left\{\begin{matrix}(BC)\\(AC)\end{matrix}\right.$		$\left.\begin{matrix}(AB)\\(CB)\end{matrix}\right\}$	=	$\left\{\begin{matrix}(CB)\\(AB)\end{matrix}\right.$		$\left.\begin{matrix}(BA)\\(CA)\end{matrix}\right\}$	=	$\left\{\begin{matrix}(CA)\\(BA)\end{matrix}\right.$
(ABC)	=	(BAC)	=	(ACB)	=	(CAB)	=	(BCA)	=	(CBA)

At this point it is worth while to put the preliminary calculations in Exercise 1 of Chapter 11 and the results in (10) and (11) above together and study their relationships. For example, the total *ssq* for the seven contrasts in (10) is

$$U - C = 7076 - 6000 = 1076, \quad \text{between cells.}$$

In (11),

$$U - B = \sum w_j d_j^2 = 626.417, \text{ treatments and interactions}$$

Also, the bottom two lines of (11) show that $(\sum w_j)\bar{d}^2 = 363.265$ is the *ssq* due to contrast *c*, adjusted for blocks; and the interaction *ssq* = 263.152 is the sum of the three interaction *ssq* $((AC), (BC), (ABC))$. To summarize, the analysis of variance consists of only three relevant quantities:

	Source	*ssq*	*df*	*msq*	*F*
$U - B$	Contrast (C), adjusted for blocks	363.265	1	363.265	7.848
	Interactions $(AC), (BC), (ABC)$	263.152	3	87.717	1.895
$A - U$	Within cells	324.000	7	46.2857	
$A - B$	Within blocks	950.417	11		

The sum of squares between blocks, ignoring treatments (c_1 and c_2), has been left out of the table, so the analysis is essentially that of the sum of squares within blocks.

Exercise

With the values of cell totals Y_i and cell sizes n_i given in (9), verify the results given in (10), using the contrast coefficients of Table 12.5.

Hint: Do it yourself before checking with the following answer.

Y_i	42	62	33	16	17	36	51	43	$\sum k_i Y_i$	$\sum k_i^2 n_i$	ssq
n_i	2	2	3	1	1	2	3	1	$=L$	$=D$	$=L^2/D$
(A)	$\frac{1}{8}$	$\frac{1}{8}$	$\frac{1}{8}$	$\frac{1}{8}$	$\frac{-1}{7}$	$\frac{-1}{7}$	$\frac{-1}{7}$	$\frac{-1}{7}$	$\frac{15}{8}$	$\frac{15}{56}$	13.125
(B)	$\frac{4}{8}$	$\frac{4}{8}$	$\frac{-4}{8}$	$\frac{-4}{8}$	$\frac{4}{7}$	$\frac{4}{7}$	$\frac{-3}{7}$	$\frac{-3}{7}$	$\frac{35}{2}$	$\frac{26}{7}$	82.452
(AB)	$\frac{1}{4}$	$\frac{1}{4}$	$\frac{-1}{4}$	$\frac{-1}{4}$	$\frac{-1}{3}$	$\frac{-1}{3}$	$\frac{1}{4}$	$\frac{1}{4}$	$\frac{235}{12}$	$\frac{13}{12}$	354.006
(C)	$\frac{2}{4}$	$\frac{-2}{4}$	$\frac{1}{4}$	$\frac{-3}{4}$	$\frac{2}{3}$	$\frac{-1}{3}$	$\frac{1}{4}$	$\frac{-3}{4}$	$\frac{-407}{12}$	$\frac{19}{6}$	363.265
(AC)	$\frac{2}{7}$	$\frac{-2}{7}$	$\frac{1}{7}$	$\frac{-3}{7}$	$\frac{-8}{17}$	$\frac{4}{17}$	$\frac{-3}{17}$	$\frac{9}{17}$	$\frac{759}{119}$	$\frac{2584}{2023}$	31.849
(BC)	$\frac{3}{14}$	$\frac{-3}{14}$	$\frac{-2}{14}$	$\frac{6}{14}$	$\frac{6}{17}$	$\frac{-3}{17}$	$\frac{-2}{17}$	$\frac{6}{17}$	$\frac{795}{119}$	$\frac{93}{119}$	57.109
(ABC)	$\frac{1}{2}$	$\frac{-1}{2}$	$\frac{-1}{3}$	1	-1	$\frac{1}{2}$	$\frac{1}{3}$	-1	-30	$\frac{31}{6}$	174.194
											1 076.000

13

Dichotomized categorical data

The analysis of general categorical data (or frequency data) is beyond the scope of this monograph; there are specialized books on that subject. However, for dichotomized categorical data the method of orthogonal contrasts described in the last chapter is also applicable. As data of this nature arise frequently in epidemiological and biomedical research, it would be well to include an application of the contrast method to categorical data. In the interest of brevity, we shall give only one example of three factors, as the case for two factors follows easily. Let the three factors be

a = social economic class

b = blood pressure

c = cholesterol level in serum

Each of these factors may be classified as 'high' or 'low' according to certain social and clinical conventions. The classification is necessarily somewhat arbitrary, as these factors are obviously continuous. Here, we shall accept the classifications as reported.

Physicians and social scientists are interested to know the effects of these three factors on the occurrence of, say, coronary heart disease (CHD). One reported result of such a study is reproduced in Table 13.1, purely for the purpose of illustrating the statistical procedure of analysis; and we shall not discuss the merit of such studies. (Recent research shows that cholesterol may be subdivided into various fractions, some of which may be related to coronary heart disease, some of which may not or, indeed, may even be protective.)

The entries in Table 13.1 are numbers of white male individuals, not measurements of a quantitative trait. The variable analogous to measurement is the presence or absence of coronary heart disease in an individual.

The nature of the data results from the convention that the presence of CHD takes the value 1 and its absence takes the value 0. Thus, in the first group of Table 13.1, there are $n_1 = 33$ individuals with values

$$0, 0, \ldots, 0; \qquad 1, 1, \ldots, 1$$

of which 26 are zeroes and 7 are ones. Thus, the number of CHD individuals in a cell is also the cell total, designated by Y_i in Table 13.1, as in the last chapter.

The chi square test

The overall test for the effects of the three factors (a, b, c) on the occurrence of coronary heart disease is to calculate a chi square with seven degrees of freedom for the eight cells of Table 13.1. The reader is assumed to be familiar with the chi square test, at least for simple cases. The chi square for the 2×8 table may be obtained in various ways. Since we have identified Y_i = number of CHD *individuals* in a cell, analogous to cell total for measurement data, we shall use the familiar procedure of calculating the sum of squares, as we have used in all previous chapters. Thus, the *ssq* between the eight cells of Table 13.1 is $(i = 1, \ldots, 8)$

$$\begin{aligned} ssq_U &= \sum_i \frac{Y_i^2}{n_i} - \frac{Y^2}{N} \\ &= \frac{7^2}{33} + \frac{7^2}{57} + \frac{8^2}{58} + \frac{9^2}{164} + \frac{3^2}{27} + \frac{2^2}{39} + \frac{4^2}{94} + \frac{4^2}{182} - \frac{44^2}{654} \\ &= 1.6756 \end{aligned} \qquad (1)$$

Table 13.1. *Prevalence of coronary heart disease (CHD) among white males aged* 40–74 *by social class, cholesterol level, and blood pressure* (J. R. McDonough, C. G. Hames, S. C. Stulb, & G. E. Garrison, 1965 *J. Chronic Diseases* **18**, 443–68)

a = social class	high				low				
c = cholesterol	high		low		high		low		
b = blood pressure	high	low	high	low	high	low	high	low	Total
No CHD	26	50	50	155	24	37	90	178	610
CHD, Y_i	7	7	8	9	3	2	4	4	$44 = Y$
Size of cell, n_i	33	57	58	164	27	39	94	182	$654 = N$

As in the theory of analysis of variance, a common error variance for the cells is assumed under the null hypothesis (that is, the factors a, b, c have no effects). For the binomial variable ($y = 1, 0$), the variance is $pq = p(1-p)$, where p is the probability that $y = 1$, and $q = 1-p$ is the probability that $y = 0$. Hence, we take the overall prevalance of CHD as an estimate of p. For the data of Table 13.1, $p = Y/N = 44/654$ and the variance is

$$\text{Var.} = pq = (\tfrac{44}{654})(\tfrac{610}{654}) = 0.062\,75 \tag{2}$$

This is equivalent to taking the total sum of squares divided by the total number (N) of individuals as the estimate of the error variance. For a variable $y = 1, 0$, we have thc simplifying situation $\sum y^2 = \sum y = Y$, where the summation covers all N individuals. Thus, the total sum of squares is

$$ssq_{\text{T}} = \sum y^2 - \frac{Y^2}{N} = Y - \frac{Y^2}{N}$$

and

$$\text{Var.} = \frac{ssq_{\text{T}}}{N} = \frac{Y}{N} - \frac{Y^2}{N^2} = p(1-p) \tag{2'}$$

Note that the estimate of the error variance here is different from that in the analysis of variance. But, when the number of observations within the cells is very large, the difference begins to disappear under the null hypothesis.

The chi square is defined as the *ssq between cells* divided by error variance:

$$\chi^2 = \frac{ssq_U}{\text{var.}} = \frac{1.6756}{0.062\,75} = 26.70 \quad \text{with 7 } df \tag{3}$$

From a chi square table, we see that this value exceeds the tabulated value at the P = 0.001 level. So, the heterogeneity of the eight cells is highly significant. In other words, the three factors (a, b, c) do effect the occurrence of CHD. But this overall test does not identify the effect of each factor separately. The analysis to be made is to test the effect of one factor specifically, following the general procedure of the last chapter.

Method of orthogonal contrasts

The gist of the contrast method is to obtain an orthogonal subdivision of the sum of squares between the cells ($ssq_U = 1.6756$) by using the coefficients of Table 12.5. The reader may well review Chapter

12, or at least its section on the analysis of three factors, before he follows the arithmetic of this section. The meaning of the notation employed in Table 12.5 must be precisely understood. For instance, $N_{12}=n_1+n_2$, and $W_{12}=w_1+w_2=(1/n_1)+(1/n_2)$.

The orthogonal subdivision of the sum of squares between the cells makes use of only the cell totals (Y_i) and cell sizes (n_i) as indicated in Table 12.5 and the numerical example given at the end of Chapter 12 (Exercise). Now the categorical data of Table 13.1 give us the value of Y_i and n_i and hence the application of the contrast method to these data is straightforward; it is merely a matter of calculating the coefficients for each contrast and then calculating the individual sum of squares by (6), (7), (8) of the last chapter. The contrast analysis for the categorical data is given in Table 13.2.

As pointed out in the last chapter, there are six possible sequences of classification with respect to three factors. The last factor in the sequence of classification is the one under investigation, while the other two factors serve as 'blocks'. In Table 13.2 the sequence of classification is *a*, *c*, *b*, so that the analysis is to obtain the *ssq* due to *b* (blood pressure) adjusted to blocks (*a*, social class, and *c*, cholesterol level). If we wish to test the significance of *c*, the sequence of classification should be *a*, *b*, *c* or *b*, *a*, *c*, by rearranging the cells.

A few words about the arithmetic involved may help the reader to follow the results of Table 13.2. With an ordinary pocket calculator it is advisable to first tabulate the values of n_i and $w_i=1/n_i$ and their partial sums before calculating the contrast coefficients. Thus,

$$n_i: \quad \underbrace{\underbrace{33 \quad 57}_{N_{12}=90} \quad \underbrace{58 \quad 164}_{N_{34}=222}}_{N_{1234}=312} \quad \underbrace{\underbrace{27 \quad 39}_{N_{56}=66} \quad \underbrace{94 \quad 182}_{N_{78}=276}}_{N_{5678}=342} \tag{4}$$

Now consider the first contrast in Table 12.5. The first four coefficients and the second four coefficients are, respectively,

$$\frac{1}{N_{1234}}=\frac{1}{312}=0.003\,205, \qquad \frac{-1}{N_{5678}}=\frac{-1}{342}=-0.002\,924$$

These numbers are inconvenient to work with, as their squares would be out of range of an ordinary pocket calculator. It was pointed out in the last chapter that the coefficients of any row may be multiplied by a constant without destroying the orthogonality of the contrasts or affecting the final value of the sum of squares. Hence, in Table 13.2, we use the coefficients 3.205 and −2.924 instead. There are other simplifying

Table 13.2. *Orthogonal subdivision of the sum of squares between cells. Contrast coefficients in Table* 12.5. *Data in Table* 13.1

a	high				low						
c	high		low		high		low				
b	high	low	high	low	high	low	high	low			
Y_i	7	7	8	9	3	2	4	4	$\sum k_i Y_i$	$\sum k_i^2 n_i$	ssq
n_i	33	57	58	164	27	39	94	182	L	D	L^2/D
(A)	3.205	3.205	3.205	3.205	−2.924	−2.924	−2.924	−2.924	61.34	6128	0.6140
(C)	7.115	7.115	−2.885	−2.885	8.070	8.070	−1.930	−1.930	75.49	11 730	0.4858
(AC)	1.111	1.111	−0.450	−0.450	−1.515	−1.515	0.362	0.362	3.22	344	0.0302
(B)	6.333	−3.667	7.387	−2.613	5.909	−4.091	6.594	−3.406	76.55	14 169	0.4136
(AB)	9.935	−5.752	11.588	−4.099	−7.582	5.249	−8.460	4.370	56.49	28 517	0.1119
(CB)	4.257	−2.465	−2.422	0.857	4.699	−3.253	−1.350	0.697	5.86	2674	0.0128
(ACB)	3.030	−1.754	−1.724	0.610	−3.704	2.564	1.064	−0.549	3.30	1500	0.0073
											$ssq_U = 1.6756$

methods, such as multiplying the first row of coefficients by (312)(342), yielding the coefficients 342, −312, or 1.71, −1.56, without rounding off errors. But we will use the form in Table 12.5 for the convenience of following the arithmetic. The coefficients in other rows have also been similarly adjusted. The following reciprocals have been multiplied by 100 for convenience:

$$w_i:\quad \underbrace{\underbrace{3.0303\quad 1.7544}_{W_{12}=4.7847}\quad \underbrace{1.7241\quad 0.6098}_{W_{34}=2.3339}}_{W_{1234}=7.1186}\quad \underbrace{\underbrace{3.7037\quad 2.5641}_{W_{56}=6.2678}\quad \underbrace{1.0638\quad 0.5495}_{W_{78}=1.6133}}_{W_{5678}=7.8811} \tag{5}$$

Substituting the values of (4) and (5) in Table 12.5, and adjusting the decimal places, we obtain the contrasting coefficients shown in Table 13.2. The actual computation was based on six significant figures in order to verify the relationship that the seven components of *ssq* add up to the *ssq* between cells.

Once the coefficients are obtained, the next step is to calculate, for each contrast, the following three quantities:

$$L=\sum k_iY_i,\quad D=\sum k_i^2n_i,\quad ssq=L^2/D \tag{6}$$

where k_i, $i=1,\ldots,8$, are the eight coefficients of a row. From the last column of Table 13.2 we see that the seven orthogonal components of *ssq* add up to $ssq_U=1.6756$, in agreement with (1).

Again, analogous to the situation in analysis of variance, there will be no test for the first three contrasts ((A), (C), (AC)), as they ignore factor b. The sum of the first three *ssq*s is the sum of squares between the four blocks (a_1c_1, a_1c_2, a_2c_1, a_2c_2), ignoring 'treatment' b. The remaining four contrasts, however, may be tested, as they have been adjusted for the other two factors. Dividing each *ssq* by the error variance $pq=0.062\,75$, we obtain a chi square with $df=1$. Thus, from the lower portion of the last column of Table 13.2,

Contrast	*ssq*	χ^2	
(B)	0.4136	6.591	
(AB)	0.1119	1.783	
(CB)	0.0128	0.205	$\chi^2=2.104$ with 3 *df*. (for AB, CB, ACB)
(ACB)	0.0073	0.116	

$\text{Var}=pq=0.062\,75$ (7)

The conclusion is that blood pressure (factor b) has a significant ($P < 0.02$) effect on the occurrence of CHD and its interactions with cholesterol and social class are negligible.

Pooling the interactions

From Table 13.2 we have

Interaction:	(AB)	(CB)	(ACB)	All interactions	
ssq:	0.1119	0.0128	0.0073	0.1320	(8)
df:	1	1	1	3	

For factors without strong interactions it is probably unnecessary to isolate each specific type of interaction. If we only wish to have a pooled *ssq* for the three types of interactions, then the shortcut method of comparing two treatments described in the last chapter, under the same section heading 'pooling the interactions', is readily applicable. The numerical results are given in Table 13.3.

Since we are primarily comparing the effects of high and low blood pressure on the occurrence of CHD, the four classifications with respect to factors a and c will serve as four 'blocks' while high b and low b serve as the two treatments, as indicated in Table 13.3. The reader should be familiar with the rest of the arithmetic, since the method has been used several times previously.

In the first cell there are $n_1 = 33$ individuals, of which $Y_1 = 7$ have CHD. The mean of this cell is $\bar{y}_1 = 7/33 = 0.212\,121 = p_1$, the proportion with CHD. However, in Table 13.3, it is given as 7 (33) to conform with our previous notation that the number of observations in a cell is given in parentheses.

The difference d_j is the difference between the cell means in block j. Thus, $d_1 = 0.212\,121 - 0.122\,807 = 0.089\,314$. We retained more significant figures than necessary in practice in order to verify certain relationships.

In order to avoid any possible confusion, we had better also point out that the *weight* w_j is not the simple reciprocal $w_i = 1/n_i$ used in the contrast coefficients (although they are related). The weight w_1 for d_1 in the first block is

$$w_1 = \frac{33 \times 57}{33 + 57} = \frac{1881}{90} = 20.90$$

Table 13.3. *Shortcut method of comparing two treatments (two levels of blood pressure) adjusted for other factors (cholesterol and social class). Cell size is in parentheses; d_j and w_j defined on page 96.*

a = social class		high		low		
c = cholesterol		high	low	high	low	
		I	II	III	IV	
b = blood pressure	high	7 (33)	8 (58)	2 (27)	4 (94)	
	$\bar{y}_{1j}$	0.212 121	0.137 931	0.111 111	0.042 553	
	low	7 (57)	9 (164)	2 (39)	4 (182)	
	$\bar{y}_{2j}$	0.122 807	0.054 878	0.051 282	0.021 978	
Difference	d_j	0.089 314	0.083 053	0.059 829	0.020 575	$\bar{d}$ = 0.054 028
Weight	w_j	20.9000	42.8468	15.9545	61.9855	$\sum w_j$ = 141.687
	$w_j d_j$	1.866 66	3.558 56	0.954 54	1.275 35	$\sum w_j d_j$ = 7.6551
	$w_j d_j^2$	0.166 72	0.295 55	0.057 11	0.026 24	$\sum w_j d_j^2$ = 0.5456
						$(\sum w_j)\bar{d}^2$ = 0.4136
					Subtracting (interactions)	= 0.1320

The other weights are calculated the same way. Each numerator is the product and each denominator is the sum of the two *n*s of the cells in the same block.

The average difference for the four blocks is

$$\bar{d} = \frac{\sum w_j d_j}{\sum w_j} = \frac{7.6551}{141.687} = 0.054\,028$$

The value of $\sum w_j d_j^2 = 0.5456$ is the *ssq* for the main effect of factor *b* *and* its interactions with the other two factors (*a* and *c*). The *ssq* due to the main effect of factor *b* only ($df = 1$) is

$$(\sum w_j)\bar{d}^2 = \frac{(\sum w_j d_j)^2}{\sum w_j} = 0.4136$$

in agreement with the *ssq* given in the last column of Table 13.2 corresponding to contrast (*B*). Finally, the pooled interactions *ssq* is obtained by subtraction:

$$\sum w_j d_j^2 - (\sum w_j)\bar{d}^2 = 0.5456 - 0.4136 = 0.1320$$

which is precisely the pooled *ssq* for the three types of interactions given in (8).

Dividing these *ssq* by the common error variance $pq = 0.062\,75$, we obtain the chi square indicated in (7).

Other methods

In this chapter we have subjected the dichotomized categorical data to the same analysis as those adopted for the measurement data in the analysis of variance, except that the chi square replaced the variance-ratio. This is equivalent to assuming the same linear models for the binomial variable $y = 1, 0$ as for the general $y_{ij\alpha}$. This assumption may or may not be justified, depending on the nature of *y* as well as the nature of the factors *a*, *b*, *c*. We shall not attempt generalizations. There exist other plausible assumptions (models) about categorical data and, hence, there are other methods of analysis. Each set of assumptions leads to a different analysis. Therefore, it is important that the reader realizes that we are not promoting any one method at the expense of others. A statistical analysis is not a routine procedure of calculations; it must pay due consideration to the subject matter such as the nature of the variable under consideration and the nature of the factors influencing it.

As a footnote we may add that the data in Table 13.1 has been analyzed by other methods (such as the ratio of cross-products) by the

authors (McDonough, *et al.*, 1965). The happy incidence is that their conclusions and ours are essentially the same, *namely*, blood pressure and cholesterol act independently on the occurrence of coronary heart disease with negligible interactions.

Exercises

1. Classify the factors in the sequence of a, b, c to test the effect of cholesterol on CHD by rearranging the cells of Table 13.1. Does the error variance $pq = 0.062\,75$ remain the same as before?

a = social class	high				low				
b = blood pressure	high		low		high		low		
c = cholesterol	high	low	high	low	high	low	high	low	Total
Number of CHD, Y_i	7	8	7	9	3	4	2	4	$Y = 44$
Size of cell, n_i	33	58	57	164	27	94	39	182	$N = 654$

Compare the present results with those for the sequence a, c, b in Table 13.2.

In Table 13.2		*Present case*		
Contrast	*ssq*	*Contrast*	*ssq*	*Chi square*
(A)	0.6140	(A)	0.6140	
(C)	0.4858	(B)	0.4898	(no test)
(AC)	0.0302	(AB)	0.1347	
(B)	0.4136	(C)	0.3987	$\chi^2 = 6.35 \quad df = 1$
(AB)	0.1119	(AC)	0.0183	
(CB)	0.0128	(BC)	0.0128	$\chi^2 = 0.61 \quad df = 3$
(ACB)	0.0073	(ABC)	0.0073	
ssq_U	1.6756	ssq_U	1.6756	

That the *ssq* for the contrast (A) in both cases remains the same is obvious. As pointed out in the last chapter, the three-factor interaction remains the same for all six sequences of classification. It should also be noted that (CB) in Table 13.2 and (BC) in the present case have the same *ssq*, as the interchange of b and c has no effect on the contrast coefficients.

2. Pool the high and low social classes of the table in Exercise 1 and test the effect of cholesterol adjusted for blood pressure. Since there will be only four classes, consult Table 12.3 for the contrast coefficients

b = blood pressure	high		low		
c = cholesterol	high	low	high	low	Total
Number of CHD, Y_i	10	12	9	13	44
Size of cell, n_i	60	152	96	346	654

Partial Answer:

	$ssq = L^2/D$	χ^2
Blood pressure, ignoring cholesterol	0.4178	6.66 (no test)
Cholesterol, adjusted for blood pressure	0.5410	8.62
Interactions	0.0272	0.43
Total	0.9860	15.71

Reference

Li, C. C. & Mazumdar, S. (1976). Analysis of dichotomized factorial data, *Journal of Chronic Disease*, **29**, 355–70.

Appendix A

Correspondence between a linear restriction and a generalized inverse

This appendix should be read together with Chapter 7, wherein the matter of introducing a linear constraint ($\sum a_i t_i = 0$) was discussed. As modern mathematical textbooks on the analysis of linear models use matrix algebra extensively, employing a 'generalized inverse' to solve the normal equations, it would be instructive to establish some connection between a linear restriction and a generalized inverse, thus bringing these two methods closer together.

An understanding of the relationship to be established in the following paragraphs requires some elementary knowledge of matrix algebra and its applications in linear model analysis. The necessary background knowledge may be gained by consulting Searle's (1966) excellent introductory text which combines matrix algebra with its statistical applications. I hope a reader with the bare minimum knowledge of matrix algebra will be able to follow the appendix.

As adopted in the text, we shall continue to explain the relationships by way of examples rather than in abstract generality. A convenient starting point is the following set of equations, obtained after eliminating the bs from the original normal equations (Chapter 7(1)):

$$\left.\begin{aligned} q_1 &= \tfrac{13}{6}t_1 - \tfrac{5}{6}t_2 - \tfrac{4}{3}t_3 = -101 \\ q_2 &= -\tfrac{5}{6}t_1 + \tfrac{23}{12}t_2 - \tfrac{13}{12}t_3 = 58 \\ q_3 &= -\tfrac{4}{3}t_1 - \tfrac{13}{12}t_2 + \tfrac{29}{12}t_3 = 43 \end{aligned}\right\} \qquad (1)$$

Since only two of the three equations are independent, we have introduced an arbitrary constraint in order to obtain a set of numerical solutions for the ts. This is what we did in Chapter 7 which the reader

may wish to review. It was also pointed out in that chapter that the constraint must be a non-estimable function of the ts.

Now let us consider the matrix of equations (1), that is, the matrix of the coefficients of the ts:

$$A = \begin{pmatrix} \frac{13}{6} & -\frac{5}{6} & -\frac{4}{3} \\ -\frac{5}{6} & \frac{23}{12} & -\frac{13}{12} \\ -\frac{4}{3} & -\frac{13}{12} & \frac{29}{12} \end{pmatrix} \qquad (2)$$

Further, write

$$t = \begin{pmatrix} t_1 \\ t_2 \\ t_3 \end{pmatrix} \text{ and } q = \begin{pmatrix} q_1 \\ q_2 \\ q_3 \end{pmatrix} = \begin{pmatrix} -101 \\ 58 \\ 43 \end{pmatrix} \qquad (3)$$

With this notation, the equations (1) may be written

$$At = q \qquad (1')$$

If matrix A were non-singular, it would have a true unique inverse A^{-1}, and there would be a unique solution for t given by $t = A^{-1}q$. Unfortunately, this is not the case here. Our present matrix A is singular with rank $r(A) = 3 - 1 = 2$. The true inverse A^{-1} does not exist; and hence there is no unique solution of the type $t = A^{-1}q$.

Generalized inverses

Then, how are we to obtain any solution for the ts at all without introducing a linear restriction on the ts? A recent (historically speaking) development in matrix algebra is the discovery of the generalized inverses of singular matrices. A singular matrix, though without a true inverse, has an infinite number of generalized inverses. A generalized inverse (of a singular matrix) is an inverse of a sort and plays certain roles analogous to those of a true inverse.

We shall use G (more convenient than A^-) to denote a generalized inverse of the singular matrix A. Analogous to $AA^{-1}A = A$ for non-singular matrices and true inverses, a generalized inverse G of a singular matrix is defined as a matrix that satisfies the condition

$$AGA = A \qquad (4)$$

This matrix G is not unique; there are an infinite number of Gs that satisfy the condition (4). Hence we always say *a* generalized inverse, not *the* generalized inverse.

Not only are there many Gs that satisfy the condition (4), but there are many different ways of finding such Gs. However, once a generalized

inverse is found by whatever method, we will obtain a particular solution for the ts given by

$$t = Gq \tag{5}$$

where G plays the role of A^{-1} in the unique solution $t = A^{-1}q$ for non-singular A. Since G is not unique, neither is the solution (5); it is a particular solution. Each particular G yields a particular solution for t, just as each linear restriction yields a particular solution for t in Chapter 7.

The correspondence

Now we come to the main point of the appendix, *namely*, to establish a connection between a linear restriction and a generalized inverse. Specifically, for each linear restriction which yields a particular solution for the ts, we wish to find a generalized inverse that yields the same solution for the ts. In other words, we wish to establish the correspondence between the equation $\sum a_i t_i = 0$ and the matrix G. The method of accomplishing this is straightforward, and may be explained in the following three steps.

As an example, let us consider restriction (vi) in Chapter 7(2), p. 61

$$4t_1 - 3t_2 + 2t_3 = 0 \tag{6}$$

The first step is to replace one of the equations in (1) by (6); say, the last equation. Then the matrix of this new set of equations is

$$A_* = \begin{pmatrix} \frac{13}{6} & \frac{-5}{6} & \frac{-4}{3} \\ \frac{-5}{6} & \frac{23}{12} & \frac{-13}{12} \\ 4 & -3 & 2 \end{pmatrix} \tag{7}$$

The matrix A_* is non-singular. The second step is to find the (true) inverse of A_*.

$$A_*^{-1} = \begin{pmatrix} \frac{14}{249} & \frac{136}{249} & \frac{1}{3} \\ \frac{-64}{249} & \frac{232}{249} & \frac{1}{3} \\ \frac{-124}{249} & \frac{76}{249} & \frac{1}{3} \end{pmatrix} \tag{8}$$

The third and final step is to replace the third column of (8) by zeros

to obtain the desired generalized inverse:

$$G = \frac{1}{249}\begin{pmatrix} 14 & 136 & 0 \\ -64 & 232 & 0 \\ -124 & 76 & 0 \end{pmatrix} \tag{9}$$

That this G is a generalized inverse of the singular matrix A in (2) may be verified by definition (4); $AGA = A$. The particular solution for the ts is

$$t = Gq = \frac{1}{249}\begin{pmatrix} 14 & 136 & 0 \\ -64 & 232 & 0 \\ -124 & 76 & 0 \end{pmatrix}\begin{pmatrix} -101 \\ 58 \\ 43 \end{pmatrix} = \begin{pmatrix} 26 \\ 80 \\ 68 \end{pmatrix} \tag{10}$$

which is precisely the solution shown in Table 7.1 under restriction (vi).

The simple procedure described above is a general one. The reader may try any one of the other restrictions employed in Chapter 7 and find the G that yields the same solution for the ts under that restriction.

In order to establish the reverse correspondence, it may be shown that the rows of G in (9) obey the same restriction (6) of the ts. Thus, for the elements of the first and second columns of (9), we observe

$$4(14) - 3(-64) + 2(-124) = 0$$
$$4(136) - 3(232) + 2(76) = 0$$

Thus, when given a G of the form (9), we can always solve for the coefficients a_i of the corresponding restriction $a_1t_1 + a_2t_2 + a_3t_3 = 0$, which yields the same solution as the G does. In solving for the coefficients a_i, it should be remembered that only their ratio $(a_1 : a_2 : a_3)$ is relevant. To summarize, we have established the two-way correspondence between a linear restriction and a generalized inverse. When one is given, we can find the other. This type of G is not just any G but a G with certain properties.

The general solution

The solution (5), $t = Gq$, is a particular solution. The general solution as given in textbooks, is

$$t = Gq + (H - I)z \tag{11}$$

where I is the identity matrix, z is an arbitrary column (of arbitrary elements), and

$$H = GA = \frac{1}{3}\begin{pmatrix} -1 & 3 & -2 \\ -4 & 6 & -2 \\ -4 & 3 & 1 \end{pmatrix} \tag{12}$$

Hence,

$$(H-I)z=\frac{1}{3}\begin{pmatrix}-4 & 3 & -2\\ -4 & 3 & -2\\ -4 & 3 & -2\end{pmatrix}\begin{pmatrix}z_1\\ z_2\\ z_3\end{pmatrix}=\begin{pmatrix}c\\ c\\ c\end{pmatrix} \tag{13}$$

$H-I$ has identical rows; consequently, $(H-I)z$ has identical elements c, a constant. This means that the solutions obtained under various restrictions or by various generalized inverses differ only by a constant, as we have shown by elementary means in Chapter 7.

The unique generalized inverse

A generalized inverse G satisfies the condition $AGA=A$ only. There exists a special generalized inverse that satisfies more conditions. Let K be a generalized inverse of a singular matrix A that satisfies the following four conditions:

$$\begin{aligned} AKA=A, &\quad (KA)'=KA \\ KAK=K, &\quad (AK)'=AK \end{aligned} \tag{14}$$

where the prime indicates transpose. The two conditions on the right mean that KA and AK are symmetrical. This K matrix, also known as the Penrose inverse, is unique and exists even for rectangular matrices.

Again, there are many methods of finding the unique generalized inverse. It is beyond the scope of the appendix to deal with such methods. For the matrix A in (2), it has been found that the unique generalized inverse is

$$K=\begin{pmatrix}1 & 0\\ 0 & 1\\ -1 & -1\end{pmatrix}\frac{2}{249}\begin{pmatrix}26 & -16 & -10\\ -16 & 29 & -13\end{pmatrix}$$

$$=\frac{2}{249}\begin{pmatrix}26 & -16 & -10\\ -16 & 29 & -13\\ -10 & -13 & 23\end{pmatrix} \tag{15}$$

yielding the solution

$$t=Kq=K\begin{pmatrix}-101\\ 58\\ 43\end{pmatrix}=\begin{pmatrix}-32\\ 22\\ 10\end{pmatrix} \tag{16}$$

which is identical with the solution shown in Table 7.1 under restriction (ii), $t_1+t_2+t_3=0$. This result was expected as soon as K was found.

Note that the elements of each column of (15) add up to zero; hence the resulting ts in $t = Kq$ will also add up to zero. It may be shown that this is a general feature of the unique generalized inverse K.

Before the advent of the generalized inverse, most statisticians used the restriction $\sum t_i = 0$ in the analysis of variance. Now we see using that restriction is equivalent to using the unique generalized inverse. However, K is not the only generalized inverse that yields $\sum t_i = 0$. If we proceed according to the three steps in (7), (8), (9), replacing the last row of (7) by (1, 1, 1), we will have obtained a G-inverse that also yields $\sum t_i = 0$. Thus,

$$t = Gq = \frac{1}{83}\begin{pmatrix} 24 & -4 & 0 \\ -2 & 28 & 0 \\ -22 & -24 & 0 \end{pmatrix}\begin{pmatrix} -101 \\ 58 \\ 43 \end{pmatrix} = \begin{pmatrix} -32 \\ 22 \\ 10 \end{pmatrix} \tag{17}$$

Solving the normal equations

There are eight normal equations (Table 4.2), involving the eight unknowns: m, b_1, b_2, b_3, b_4 and t_1, t_2, t_3. The matrix method enables us to solve the set of normal equations at once. With eight equations and eight unknowns, we need to deal with 8×8 matrices. The matrix method becomes practical only when high speed computers are available; otherwise, it actually involves more arithmetic than the elementary method. The chief advantage of the matrix method lies in its generality and succinctness in theoretical work, not in arithmetic economy.

In the preceding simple case the rank of the matrix is $r(A) = 3 - 1$. Now we wish to extend our method to the more general situation where the rank of the matrix of the normal equations is $r = 8 - 2$. We continue to use the same numerical example as in Chapters 4–7. Each observed y is expressed as a sum of m, b, t, and e, as in the upper portion of Table 4.2, and there are $N = 11$ such observed single values. In matrix notation, these eleven equations are summarized as

$$y = Xb + e \qquad \text{(linear model)}$$

where y is a column of the N observed single values; X is an 11×8 matrix consisting of elements 1, or 0, depending on the presence or absence of a parameter in a single observed value; b is a column of eight unknown parameters (m, b_1, b_2, b_3, b_4, t_1, t_2, t_3); and e is a column of N residual values (error). The matrix X may be called the occurrence matrix (or the design matrix), because it shows where the observations are located.

The normal equations corresponding to the linear model shown above always take the form:

$$(X'X)b = X'y \tag{18'}$$

In our particular numerical example, the eight normal equations (Table 4.2) are

$$\left(\begin{array}{c|cccc|ccc} 11 & 4 & 2 & 3 & 2 & 4 & 3 & 4 \\ \hline 4 & 4 & 0 & 0 & 0 & 2 & 1 & 1 \\ 2 & 0 & 2 & 0 & 0 & 1 & 0 & 1 \\ 3 & 0 & 0 & 3 & 0 & 1 & 1 & 1 \\ 2 & 0 & 0 & 0 & 2 & 0 & 1 & 1 \\ \hline 4 & 2 & 1 & 1 & 0 & 4 & 0 & 0 \\ 3 & 1 & 0 & 1 & 1 & 0 & 3 & 0 \\ 4 & 1 & 1 & 1 & 1 & 0 & 0 & 4 \end{array}\right)\left(\begin{array}{c} m \\ \hline b_1 \\ b_2 \\ b_3 \\ b_4 \\ \hline t_1 \\ t_2 \\ t_3 \end{array}\right)=\left(\begin{array}{c} Y \\ \hline Y_1 \\ Y_2 \\ Y_3 \\ Y_4 \\ \hline Z_1 \\ Z_2 \\ Z_3 \end{array}\right)=\left(\begin{array}{r} 550 \\ \hline 200 \\ 46 \\ 192 \\ 112 \\ \hline 86 \\ 228 \\ 236 \end{array}\right) \tag{18}$$

where $X'X$ is the 8×8 matrix on the left; b is the column of eight unknowns; and $X'y$ is the column of eight observed totals (grand total, four block totals, and three treatment totals).

Comparing (18′) with (1′), we see that $X'X$ takes the place of A; b takes the place of t; and $X'y$ takes the place of q.

The matrix $X'X$ of (18) is not of full rank. The four rows corresponding to the four blocks and the last three rows corresponding to the three treatments both add up to the first row. Hence, the rank of $X'X$ is $r = 8 - 2 = 6$.

To find a generalized inverse of $X'X$, we may follow the same three steps exemplified in (7), (8), (9), using, say, the two restrictions

$$b_1 + b_2 + b_3 + b_4 = 0, \quad t_1 + t_2 + t_3 = 0 \tag{19}$$

as we did in Chapter 7(14). The first step is to obtain the non-singular matrix $(X'X)_*$ by inserting the restrictions (19) onto $X'X$, that is,

replacing its fifth row by (0, 1, 1, 1, 1, 0, 0, 0)

and

replacing its last row by (0, 0, 0, 0, 0, 1, 1, 1)

The second step is to obtain the (true) inverse $(X'X)_*^{-1}$. The third and final step is to obtain the generalized inverse by replacing the fifth and last columns of $(X'X)_*^{-1}$ by two columns of 0's. The generalized inverse

so obtained is

$$G=\frac{1}{1992}\left(\begin{array}{r|rrrr|rrr}
210 & -120 & 24 & -80 & 0 & 30 & 78 & 0 \\
\hline
-222 & 696 & 60 & 132 & 0 & -174 & -54 & 0 \\
-378 & 216 & 1152 & 144 & 0 & -54 & 258 & 0 \\
-210 & 120 & -24 & 744 & 0 & -30 & -78 & 0 \\
810 & -1032 & -1188 & -1020 & 0 & 258 & -126 & 0 \\
\hline
48 & -312 & -336 & -208 & 0 & 576 & -96 & 0 \\
-336 & 192 & 360 & 128 & 0 & -48 & 672 & 0 \\
288 & 120 & -24 & 80 & 0 & -528 & -576 & 0
\end{array}\right) \tag{20}$$

Similar to G in (9), the rows of (20) obey the same restrictions as (19). The four rows corresponding to the four blocks add up to $(0, \ldots, 0)$ and the last three rows corresponding to the three treatments also add up to $(0, \ldots, 0)$. The solution is:

$$b=G(X'y) \tag{21'}$$

where, in our numerical case,

$$\begin{aligned} b' &= (m, b_1, b_2, b_3, b_4, t_1, t_2, t_3) \\ &= (49, 9, -15, 15, -9, -32, 22, 10) \end{aligned} \tag{21}$$

in agreement with solution (15) of Chapter 7.

Another type of restrictions

In Exercise 1 at the end of Chapter 7, we have studied the consequences of the two simultaneous restrictions

$$m=0, \quad b_1=0 \tag{22}$$

Now we wish to show that the method of finding a generalized inverse corresponding to given restrictions, described above, also applies to this case. To form $(X'X)_*$ from $X'X$ of the normal equations of (18), we replace the first two rows of $X'X$ by

$$\begin{pmatrix} 1, 0, 0, 0, 0, 0, 0, 0 \\ 0, 1, 0, 0, 0, 0, 0, 0 \end{pmatrix}$$

so that the non-singular matrix is

$$(X'X)_* = \left(\begin{array}{c|c} I & 0 \\ \hline R & M \end{array}\right) \quad \text{with } (X'X)_*^{-1} = \left(\begin{array}{c|c} I & 0 \\ \hline S & M^{-1} \end{array}\right) \tag{23}$$

where I is the 2×2 identity matrix, and M is the 6×6 principal minor matrix in the lower right. R and S do not matter too much, as the first two columns of $(X'X)_*^{-1}$ will be replaced by 0's. In other words, the generalized inverse of $X'X$ under restriction (22) is simply the true inverse M^{-1}, augmented by two rows of 0's on the top and two columns of 0's on the left, as shown below:

$$G=\frac{1}{166}\left(\begin{array}{rrrrr|rrr} 0 & 0 & 0 & 0 & 0 & 0 & 0 & 0 \\ 0 & 0 & 0 & 0 & 0 & 0 & 0 & 0 \\ 0 & 0 & 131 & 41 & 40 & -43 & -27 & -53 \\ 0 & 0 & 41 & 99 & 48 & -35 & -49 & -47 \\ 0 & 0 & 40 & 48 & 144 & -22 & -64 & -58 \\ \hline 0 & 0 & -43 & -35 & -22 & 61 & 19 & 25 \\ 0 & 0 & -27 & -49 & -64 & 19 & 93 & 35 \\ 0 & 0 & -53 & -47 & -58 & 25 & 35 & 81 \end{array}\right) \tag{24}$$

When it is multiplied on the right by the column vector $X'y$ of observed totals, the solution is

$$\begin{aligned} b' &= (m, b_1, b_2, b_3, b_4, t_1, t_2, t_3) \\ &= (0, 0, -24, 6, -18, 26, 80, 68) \end{aligned} \tag{25}$$

in complete agreement with the solutions given at the end of Chapter 7. In such a case, the situation is equivalent to solving for six unknowns from six independent equations.

The unique generalized inverse of $X'X$

The unique generalized inverse of $(X'X)$ is not frequently employed in solving the normal equations, at least not for practical analyses. It is naturally more difficult to find. The unique generalized inverse of the matrix $X'X$ of (18) has been found to be:

$$K=\left(\begin{array}{r|rrr|rr} 1 & 0 & 0 & 0 & 0 & 0 \\ \hline 0 & 1 & 0 & 0 & 0 & 0 \\ 0 & 0 & 1 & 0 & 0 & 0 \\ 0 & 0 & 0 & 1 & 0 & 0 \\ 1 & -1 & -1 & -1 & 0 & 0 \\ \hline 0 & 0 & 0 & 0 & 1 & 0 \\ 0 & 0 & 0 & 0 & 0 & 1 \\ 1 & 0 & 0 & 0 & -1 & -1 \end{array}\right)\left(\begin{array}{ccccccccc} - & - & - & - & - & - & - & - & - \\ - & - & - & - & - & - & - & - & - \\ - & - & - & - & - & - & - & - & - \\ - & - & - & - & - & - & - & - & - \\ - & - & - & - & - & - & - & - & - \\ - & - & - & - & - & - & - & - & - \end{array}\right) \tag{26}$$

$$
=\frac{1}{29963}\left(\begin{array}{r|rrrr|rrr}
1212 & -419 & 949 & -39 & 721 & 347 & 803 & 62 \\
\hline
-419 & 6844.5 & -2380 & -1544 & -3339.5 & -1628 & 291 & 918 \\
949 & -2380 & 12\,041 & -3368 & -5344 & -1533 & 3274 & -792 \\
-39 & -1544 & -3368 & 7937 & -3064 & 63 & -545 & 443 \\
721 & -3339.5 & -5344 & -3064 & 12\,468.5 & 3445 & -2217 & -507 \\
\hline
347 & -1628 & -1533 & 63 & 3445 & 6354 & -3602 & -2405 \\
803 & 291 & 3274 & -545 & -2217 & -3602 & 7380 & -2975 \\
62 & 918 & -792 & 443 & -507 & -2405 & -2975 & 5442
\end{array}\right) \qquad (26')
$$

where $29\,963 = 19 \times 19 \times 83$. The first matrix of the product form (26) is to show the structure of the unique generalized inverse. The second matrix of (26) is of order 6×8; it is not explicitly written out as it is the same as above without the 5th and 8th rows. The product form (26) is not unique but the final product is unique. In the final form of K, we note

$$
\begin{aligned}
\text{its first row} &= \text{sum of rows } 2, 3, 4, 5 \\
&= \text{sum of rows } 6, 7, 8
\end{aligned} \qquad (27)
$$

on account of the structure of the first matrix of (26). The solution given by the unique generalized inverse is $b = K(X'y)$ shown below:

$$
b = K\begin{pmatrix} 550 \\ \hline 200 \\ 46 \\ 192 \\ 112 \\ \hline 86 \\ 228 \\ 236 \end{pmatrix} = \begin{pmatrix} m \\ \hline b_1 \\ b_2 \\ b_3 \\ b_4 \\ \hline t_1 \\ t_2 \\ t_3 \end{pmatrix} = \frac{1}{19}\begin{pmatrix} 588 \\ \hline 318 \\ -138 \\ 432 \\ -24 \\ \hline -412 \\ 614 \\ 386 \end{pmatrix} \qquad (28)
$$

where

$$
m = b_1 + b_2 + b_3 + b_4 = t_1 + t_2 + t_3 \qquad (29)
$$

on account of the property (27) of the K-inverse. The solution (28) is in complete agreement with the solutions (21)–(24) of Chapter 7.

In conclusion, the appendix has shown that, for each set of linear restrictions, we can always find the corresponding generalized inverse that yields the same solutions as the restrictions do.

Variance of treatment differences

The subject of finding the variance of the difference between two treatment effects does not seem to belong to the general theme of this appendix. However, the desired variance takes the simplest form once a generalized inverse is found, and we have found several generalized inverses in the preceding sections. Hence, it is convenient to mention the subject here, making use of the generalized inverses already known.

With respect to the t-equations of (1) we have given three generalized inverses namely, (9), (15), (17). They are reproduced in the following for convenient reference, as we shall study certain invariant properties among the elements of these matrices:

$$\frac{1}{249}\begin{pmatrix} 14 & 136 & 0 \\ -64 & 232 & 0 \\ -124 & 76 & 0 \end{pmatrix}, \quad \frac{2}{249}\begin{pmatrix} 26 & -16 & -10 \\ -16 & 29 & -13 \\ -10 & -13 & 23 \end{pmatrix}, \quad \frac{1}{83}\begin{pmatrix} 24 & -4 & 0 \\ -2 & 28 & 0 \\ -22 & -24 & 0 \end{pmatrix} \tag{30}$$

In addition, with respect to the complete set of eight normal equations, we have found three 8×8 generalized inverses, namely, (20), (24), (26). Since we are dealing with treatment effects, only their lower right 3×3 submatrices are reproduced in the following:

$$\frac{1}{1992}\begin{pmatrix} 576 & -96 & 0 \\ -48 & 672 & 0 \\ -528 & -576 & 0 \end{pmatrix}, \quad \frac{1}{166}\begin{pmatrix} 61 & 19 & 25 \\ 19 & 93 & 35 \\ 25 & 35 & 81 \end{pmatrix},$$

$$\frac{1}{29\,963}\begin{pmatrix} 6354 & -3602 & -2405 \\ -3602 & 7380 & -2975 \\ -2405 & -2975 & 5442 \end{pmatrix} \tag{31}$$

Now, a few words about the general situation first. As before, let b be the column vector of the unknown parameters and let $k' = (k_1, k_2, \ldots)$ be a row of numbers such that $k'b$ is an estimable function. Then the variance of this estimable function is (Searle, 1971, p. 182, p. 279)

$$\text{Var}\,(k'b) = k'Gk\,\sigma^2 \tag{32}$$

where $\sigma^2 = E(e^2)$ is the error variance and G is the generalized inverse employed to obtain the particular solution for b. The important feature of (32) is that, although G is not unique, the quantity (scalar) $k'Gk$ is invariant to the choice of G. In other words, no matter which G we happen to use, the value of $k'Gk$ remains the same provided that $k'b$ is estimable.

Returning to our numerical example with three treatments, $t' = (t_1, t_2, t_3)$, let us find the variance of the difference $t_1 - t_2$, which is an estimable function. In this case we take $k' = (1, -1, 0)$, so that $k't = t_1 - t_2$. Let g_{ij} be the elements of G. Then, by the general theorem (32), we have

$$\text{Var}\,(t_1 - t_2) = (g_{11} - g_{12} - g_{21} + g_{22})\sigma^2 \tag{33}$$

The numerical value of (33) may be calculated from any one of the six matrices in (30) and (31). Thus, in respective order, the values of $g_{11} - g_{12} - g_{21} + g_{22}$ are as follows:

$$\frac{14 - 136 + 64 + 232}{249}, \quad \frac{2(26 + 16 + 16 + 29)}{249}, \quad \frac{24 + 4 + 2 + 28}{83}, \qquad \text{(from (30))}$$

$$\frac{576 + 96 + 48 + 672}{1992}, \quad \frac{61 - 19 - 19 + 93}{166}, \quad \frac{6354 + 2(3602) + 7380}{29\,963}, \qquad \text{(from (31))}$$

each reducing to the same value, 58/83, invariant to the choice of G.

The variance of other treatment differences may be found in the same way. For instance,

$$\text{Var}\,(t_2 - t_3) = (g_{22} - g_{23} - g_{32} + g_{33})\sigma^2$$

It is a good exercise for the reader to calculate the value of $g_{22} - g_{23} - g_{32} + g_{33}$ for each of the six matrices in (30) and (31) and check if each reduces to the same value, 52/83. Then the reader may try more complicated estimable functions such as $k't = t_1 + t_2 - 2t_3$ by taking $k' = (1, 1, -2)$. It will be found $k'Gk = 138/83$ for each of the matrices in (30) and the submatrices in (31).

References

1. Graybill, F. A. (1969). *Introduction to matrices with applications in statistics.* Wadsworth Publishing Co., Belmont, California.
2. Mazumdar, S., Li, C. C. & Bryce, G. R. (1980). Correspondence between a linear restriction and a generalized inverse in linear model analysis, *The American Statistician*, **34**(2), 103–5.
3. Searle, S. R. (1966). *Matrix algebra for the biological sciences (including applications in statistics).* John Wiley, New York and London.
4. Searle, S. R. (1971). *Linear models.* John Wiley, New York and London.

Appendix B

Independence of quadratic forms

This is an appendix to Chapter 8 on a type of orthogonal contrasts among treatments for unbalanced data. The same notation will be adopted, as frequent reference to that chapter is necessary. For convenience, the expressions of that Chapter will be referred as (8: x); for example, the equations of (6) of Chapter 8 will be referred to as (8: 6). The main purpose here is to prove that the two quadratic forms, ϕ_1 and ϕ_2, given by (8: 8) and (8: 9), are independently distributed in repeated sampling. Some repetition may be tolerated to provide self-containment and continuity in reading.

The *t*-equations

The t-equations are derived from the normal equations by eliminating the bs. It is an intermediate step in solving the normal equations. Searle (1971, p. 266) called the t-equations the 'absorbing equations', for he feels that the bs are being absorbed by these t-equations.

At the end of Chapter 4 (Exercise 3), we have already given limited generalization of the procedure for obtaining the t-equations. Let G_i be the 'size' of blocks and N_j be the total number of times that treatment j appears in the whole set of data. Also, let n_{ij} be the number of observations in block i with treatment j; that is, the size of cell (ij). Thus, $N_j = \sum_i n_{ij}$. Finally, let Z_j be the observed total of treatment j and $\bar{y}_i$ be the mean value of block i. The numerical values of the qs are:

$$q_j = Z_j - \sum_i n_{ij}\bar{y}_i \tag{1}$$

Substituting the linear model values of $\bar{y}_i$ in (1), we obtain the

t-equations:

$$q_j = \left(N_j - \sum_i \frac{n_{ij}^2}{G_i}\right) t_j - \sum_{j'} \left(\sum_i \frac{n_{ij} n_{ij'}}{G_i}\right) t_{j'} \tag{2}$$

where $j' \neq j$; that means that j' denotes all the ts other than t_j. The coefficients of the ts in (2) add up to zero, as $N_j = \sum_i n_{ij}$ and $G_i = \sum_j n_{ij}$. In the t-equations given by (8: 6), the coefficients are

$$a = \sum_i \frac{n_{i1} n_{i2}}{G_i}, \quad b = \sum_i \frac{n_{i1} n_{i3}}{G_i}, \quad c = \sum_i \frac{n_{i2} n_{i3}}{G_i} \tag{3}$$

We shall refer to these expressions later when we investigate the variance–covariance of the qs.

Treatment sum of squares

A solution for the ts from the absorbing equations (8: 6) must be found first before we can calculate the treatment *ssq* as given by the quadratic form ϕ in (8: 7). A solution for the ts may be obtained by using a generalized inverse. Let

$$A = \begin{pmatrix} a+b & -a & -b \\ -a & a+c & -c \\ -b & -c & b+c \end{pmatrix}, \quad A^- = \begin{pmatrix} a+b & a & b \\ a & a+c & c \\ b & c & b+c \end{pmatrix} \frac{1}{D} \tag{4}$$

where A^- is a generalized inverse of A, and $D = ab + ac + bc$. We have chosen A^- such that $AA^-A = A$ but $A^-AA^- \neq A^-$. Writing $t' = (t_1, t_2, t_3)$ and $q' = (q_1, q_2, q_3)$, we see that the t-equations (8: 6) and a solution may be expressed as

$$At = q, \quad t^\circ = A^- q \tag{5}$$

where the superscript $^\circ$ indicates that t° is a particular solution of the t-equations, depending on the particular A^- we choose to use. In calculating numerical values, it is always understood that we use t°. Hence, we shall omit the superscript in the following expressions for ease of printing but it is understood that it is there. The treatment *ssq* within the blocks is given by the quadratic form (8: 7).

$$\phi = t'q = q't = q'A^-q = t'At \tag{6}$$

The value of ϕ in (6) is invariant to A^- and thus invariant to t°. Hence, any set of particular solutions for the ts may be used to calculate the treatment sum of squares. It is this sum of squares that we wish to decompose into components, each with a single degree of freedom, by using orthogonal contrasts among the treatments.

The reduction or decomposition

Since the coefficients add up to zero for each of the t-equations, the first equation may be considered as a contrast between treatment (1) on one hand and treatments (2 and 3) on the other. The sum of squares due to this contrast is

$$\phi_1 = \frac{1}{a+b}[(a+b)t_1 - at_2 - bt_3]^2 = t'A_1 t \tag{7}$$

where $t = t^\circ$ and

$$A_1 = \begin{pmatrix} a+b & -a & -b \\ -a & \dfrac{a^2}{a+b} & \dfrac{ab}{a+b} \\ -b & \dfrac{ab}{a+b} & \dfrac{b^2}{a+b} \end{pmatrix} \tag{8}$$

By subtraction from (6) we obtain the remaining sum of squares:

$$\phi_2 = \phi - \phi_1 = t'A_2 t = k(t_2 - t_3)^2 \tag{9}$$

where

$$A_2 = A - A_1 = \begin{pmatrix} 0 & 0 & 0 \\ 0 & 1 & -1 \\ 0 & -1 & 1 \end{pmatrix} k \tag{10}$$

and

$$k = a + c - \frac{a^2}{a+b} = b + c - \frac{b^2}{a+b} = c + \frac{ab}{a+b}$$

The sum of squares in (9) is clearly due to the comparison $t_2 - t_3$.

To summarize, the treatment sum of squares, ϕ with two degrees of freedom, has been subdivided into ϕ_1 and ϕ_2 based on two comparisons: $(a+b)t_1 - at_2 - bt_3$ and $t_2 - t_3$, yielding

$$t'At = t'A_1 t + t'A_2 t \qquad (6) = (7) + (9)$$

where $A = A_1 + A_2$, with rank $r(A) = r(A_1) + r(A_2) = 1 + 1 = 2$. Similar subdivisions may be accomplished by using the second or third equation of (8: 6) as the first comparison.

Independence of quadratic forms

Since ϕ_2 was obtained by subtraction in the reduction process described above, it is crucial to demonstrate that ϕ_1 and ϕ_2 are independently distributed. To do this, we need to know the variance–covariance

matrix of the qs. From the solution $t^\circ = A^- q$ in (5), we have

$$E(t^\circ) = A^- E(q), \quad \text{and} \quad V_t = V(t^\circ) = A^- V(q) A^- \tag{11}$$

where E denotes expected value and V is the variance–covariance matrix of the indicated variables. Now we proceed to find $V(q)$. The values of the qs are given in (1), from which it may be readily found that

$$\begin{aligned}\text{Var}\,(q_j) &= \text{Var}\left(Z_j - \sum_i n_{ij}\bar{y}_i\right) \\ &= N_j\sigma^2 + \sum_i n_{ij}^2\left(\frac{\sigma^2}{G_i}\right) - 2\sum_i n_{ij}\left(\frac{n_{ij}\sigma^2}{G_i}\right) \\ &= \left(N_j - \sum_i \frac{n_{ij}^2}{G_i}\right)\sigma^2 \end{aligned} \tag{12}$$

where $\sigma^2 = E(e_{ij\alpha}^2) =$ error variance. The covariance between two different qs, q_j and $q_{j'}$, where $j' \neq j$, may be found in a similar way, noting $\text{Cov}(Z_j, Z_{j'}) = 0$.

$$\begin{aligned}\text{Cov}\,(q_j, q_{j'}) &= \text{Cov}\left(Z_j - \sum_i n_{ij}\bar{y}_i, Z_{j'} - \sum_i n_{ij'}\bar{y}_i\right) \\ &= 0 - \sum_i \frac{n_{ij'}n_{ij}\sigma^2}{G_i} - \sum_i \frac{n_{ij}n_{ij'}\sigma^2}{G_i} + \sum_i n_{ij}n_{ij'}\left(\frac{\sigma^2}{G_i}\right) \\ &= -\sum_i \frac{n_{ij}n_{ij'}}{G_i}\sigma^2 \end{aligned} \tag{13}$$

The coefficients of σ^2 in (12) and (13) are precisely the same coefficients of the ts in equations (2) and (3). Using matrix A in (4), we conclude

$$V(q) = A\sigma^2 \tag{14}$$

and from (11),

$$V_t = V(t) = A^- A A^- \sigma^2 \tag{15}$$

The two quadratic forms $\phi_1 = t' A_1 t$ and $\phi_2 = t' A_2 t$ given in (7, 8, 9, 10) are non-negative definite and $V_t = V(t)$ is singular. Under these circumstances, the sole condition for the two quadratic forms to be independent is (Searle, 1971, p. 71)

$$A_1 V_t A_2 = 0; \quad \text{that is, } A_1 A^- A A^- A_2 = 0 \tag{16}$$

Computation by substituting the matrices (4), (8) and (10) yields $A_1 A^- A = A_1$ and $A A^- A_2 = A_2$. Further computation confirms that

$$(A_1 A^- A) A^- A_2 = A_1 A^- (A A^- A_2) = A_1 A^- A_2 = 0 \tag{17}$$

If we write out in full, ignoring the scalars D and k and using $s = a + b$,

(17) becomes

$$\begin{pmatrix} a+b & -a & -b \\ -a & a^2/s & ab/s \\ -b & ab/s & b^2/s \end{pmatrix} \begin{pmatrix} a+b & a & b \\ a & a+c & c \\ b & c & b+c \end{pmatrix} \begin{pmatrix} 0 & 0 & 0 \\ 0 & 1 & -1 \\ 0 & -1 & 1 \end{pmatrix} = 0 \tag{18}$$

as the reader may readily verify. This completes the proof that the two quadratic forms ϕ_1 and ϕ_2 are independently distributed.

An extension

When there are four treatments, the t-equations remain $At = q$ as in (5), but the matrix A is now of the following form:

$$A = \begin{pmatrix} d+e+f & -d & -e & -f \\ -d & d+a+b & -a & -b \\ -e & -a & e+a+c & -c \\ -f & -b & -c & f+b+c \end{pmatrix} \tag{19}$$

The treatment sum of squares within the blocks is given by the quadratic form $\phi = t'At$ with three degrees of freedom. Let us use the first absorbing equation as our initial contrast. The sum of squares due to that contrast is, writing $s = d+e+f$ for brevity,

$$\phi_1 = \frac{1}{s}(st_1 - dt_2 - et_3 - ft_4)^2 = t'A_1t \tag{20}$$

where

$$A_1 = \begin{pmatrix} s & -d & -e & -f \\ -d & d^2/s & de/s & df/s \\ -e & de/s & e^2/s & ef/s \\ -f & df/s & ef/s & f^2/s \end{pmatrix} \tag{21}$$

with rank $r(A_1) = 1$. The remaining sum of squares obtained by subtraction is

$$\phi - \phi_1 = t'A_*t$$

with two degrees of freedom and rank $r(A_*) = 2$, where

$$A_* = \begin{pmatrix} 0 & 0 & 0 & 0 \\ 0 & a'+b' & -a' & -b' \\ 0 & -a' & a'+c' & -c' \\ 0 & -b' & -c' & b'+c' \end{pmatrix} \tag{22}$$

and

$$a' = a + de/s, \quad b' = b + df/s, \quad c' = c + ef/s$$

The lower right 3×3 submatrix of (22) is of the same form as the matrix A for three treatments as given in (4). Hence, the sum of squares $t'A_*t$ with two degrees of freedom may be further subdivided into two components, each with a single degree of freedom, by the method of reduction for three treatments. Thus, $A_* = A_2 + A_3$, so that $A = A_1 + A_2 + A_3$ and $\phi = \phi_1 + \phi_2 + \phi_3$. These three sums of squares are independently distributed.

An examination of the structure of the various matrices shows that the procedure of decomposing the sum of squares described here is a general one. With p treatments, the original quadratic form $t'At$ may be decomposed into one representing the contrast between one treatment (say, t_1) versus all the others and the remaining quadratic form involving only $p-1$ treatment effects. This, in turn, may be subdivided into one representing the contrast between one of the $p-1$ treatments (say, t_2) versus the other $p-2$ treatments. This continues until the last comparison (say, $t_{p-1} - t_p$). The resulting comparisons are of the type:

t_1	t_2	t_3	t_4
+	−	−	−
	+	−	−
		+	−

There are p choices at the first stage, $p-1$ choices at the second stage, etc. There are, altogether, $p(p-1)(p-2) \cdots 3 = p!/2$ ways of decomposing the treatment sum of squares. For practical purposes, however, the investigator will only be interested in one or two particular decompositions, depending on the nature and meaning of the treatments.

In closing we wish to remind the reader that this is only one type out of many orthogonal contrasts that may be made among the treatments. Comparisons for factorial types of treatments have been given in Chapters 12 and 13.

References

1. Li, C. C. & Mazumdar, S. (1981). A type of orthogonal contrasts for unbalanced data, *Biometrical Journal* (*Biometrische Zeitschrift*), 23 (7), 645–51
2. Searle, S. R. (1971). *Linear models.* John Wiley, New York and London.

Index